The Last Whale

The thud of the harpoon firing was an assault on Tom Barber's ears. The sound shocked and bruised. Time stretched like a tape recorder on slow speed. The world around him detached, was drawn out, elongated. He looked up and saw the harpoon fly towards the whale. From his viewpoint, sitting at sea level, the harpoon looked like it was coming right over his head. He couldn't understand that this was happening to him. This must be what an out-of-body experience is like, he thought. He didn't feel scared. He was in someone else's novel, a dream; he was somewhere beyond fear and death. There's nowhere to hide in a rubber boat. You may as well be sitting naked on top of a pole.

*

It had been hit. Jonny sagged. He felt impotent, useless, filled with overwhelming rage and frustration. Heavy duty. His focus was shattered, but a calm part of his mind, a splinter, noted the efficiency of the whalers. They put two more harpoons into the whale, making sure it was dead, brought it alongside, pumped it with air, put a flag on the body and disappeared over the horizon for the next kill. Blood-red waves lapped around a creature as big as a house.

'What are we going to do, Tom?'

'Let's just touch him and say goodbye.'

Screeching seagulls swarmed over bloody bits of whale as Jonny rubbed his hand along the whale's side. He took the tiller and Tom did the same, also touching the eye of the whale.

The Last Whale
Chris Pash

The Last Whale contains a reference to a member of the Indigenous community who has passed away.

GREENPEACE

A portion of all royalties
earned through sales of *The Last Whale*
is being paid to Greenpeace Australia.

One of Australia's last whalers steams into Albany, Western Australia.
Photograph by Ed Smidt.

*to the
people of Albany,
Western Australia*

The Last Whale

Harpoon.
Photograph by Ed Smidt.

1 : The Chaser : 20 June 1977

A target resembling a huge granite rock emerges from the rolling swell of the Southern Ocean. A giant tail rises at right angles to the water and slips cleanly into the sea. Foaming green water covers it over.

In a tiny alcove on the steel-hulled *Cheynes III*, Tom Kennedy, sonar operator, knows what the whale is doing. He waits for it to go deeper. It's hard to track a whale close to the surface; Tom's ears become headphones. Sound travels well in water and the pinging of the sonar bounces off solid shapes. The echoes create an image on a small screen and a jagged line on graph paper.

This whale won't get away.

Tom keeps a picture in his mind; a shadow, a guesstimate fed by experience. He is more than the science of his instruments; he *senses* the whale.

The whale knows it's being shadowed and sends its own sonar 'clicks' and 'burrs' streaming in short bursts at the ship. In Tom's mind these sounds are angry talk, swearing.

The whale dives, adjusting its body for buoyancy as the water pressure around it increases.

Sperm whales don't have vocal cords. The sounds Tom is hearing are made by the whale shifting air between cavities in its head. A sperm whale has an acoustic sense of its world. Whistles communicate to the pod; clicks are for navigating, and deep probing of the ocean looking for squid, its main diet.

The rest of the crew of seventeen are quiet, knowing Tom has a line on the whale. They scan the sea, waiting for it to surface. A big sperm whale can stay deep for thirty minutes or more.

Soon Tom will start calling distance and direction between the ship and where he thinks the whale will breach. He's sure this is a sperm whale of good size, forty tonnes or more.

A light plane buzzes above. Spotter pilot John Bell radios the coordinates of another pod.

Operating the sonar is more art than science. Tom, beginning as a deckhand, learnt his trade from an ex-navy man, Ron Day, by listening and watching. But when his chance at the sonar came, Tom was useless. The sounds piped through the headphones were indistinguishable from one another. The company sent him to get his ears checked, but they were working fine.

Tom didn't give up. It took three frustrating months, but eventually he assimilated all the information needed to track a whale underwater.

'There's a hell of a lot of concentration in it,' Tom says. 'There are a lot of other noises. Could be fish, or you could get a bounce from the bottom. We had operators who came out of the navy after tracking submarines for years. They had no hope with whales.' (Tracking the solid steel of ships or underwater topography was easy because they gave off solid images. Whales are organic, move between the surface and the depths, and give off a more ghost-like image.)

Sometimes Tom has been fooled into tracking large shapes well past the time a sperm whale should have come up for air. 'Talk about the Loch Ness monster. There are things out there — animals or fish or whatever — that put out a graph like a whale or even bigger. We've followed the sods for a couple of hours or more. What they were I don't know.'

The whales themselves know a trick or two. Once Tom chased two whales that were right there on the screen in front of him — and then they were gone. Whales can drop to a different thermal layer, a band of a higher or lower temperature than the surrounding water. This plays havoc with the sonar. Tom thinks there are fast-moving currents at different depths, like the fast lane on a highway. A whale hitting a lane like that will disappear as fast as a rabbit down a burrow.

Mick Stubbs, first mate, is resting off watch in a bunk bed. He feels a subtle change, a vibration in the ship's hull as the engines change tempo. Mick has been up since 4am, when the *Cheynes III* left Albany's Princess Royal Harbour.

As a child he went out with his father, whaling ship's master Ches Stubbs, to watch the men catch humpbacks, toothless baleen whales off Albany. It was exciting and Mick couldn't wait to go whaling himself. It was in his blood, and the pay was good. He got his start in 1961, signing

Mick Stubbs, first mate of the *Cheynes III*.
Photograph by Ed Smidt.

The Last Whale • 13

on as a twenty-year-old deckhand for the Nor' West Whaling Company at Carnarvon in the state's north-west. Back then, the horizon had been full of humpback whales. But the catches dropped to nothing the following year and humpback whaling stopped in 1963. The whales had been overfished to the point of near extinction as whaling stations on the east and west coasts of Australia took catches and with large ocean-going whaling fleets scooping up as many humpbacks, of any size or sex, as they could.

Mick ended up back in Albany working for the Cheynes Beach Whaling Company, hunting toothed sperm whales. Jobs on the whale chasers were gold in the town of Albany. You were top of the heap when you worked the chasers.

By thirty-six, Mick had made his way up from the engine room to the top deck and was second in charge of the *Cheynes III*. He had worked all corners of the ship and knew every piece of machinery. With a bit of study, and luck, he could be in line for a skipper's job in a few years.

Mick's job now is to ensure the equipment is in perfect order for when the whale they are hunting blows. The night before, he pulled the block out of the harpoon gun and inspected and cleaned the metal casing where the shells are loaded. He packed twenty-four shells with gunpowder — 185 grams each for a full shot. The fifty-five-kilogram-steel harpoon the explosive drives has flukes to hold a whale fast. Another harpoon without barbs — usually fired at close range to finish off a wounded whale — carries eighty grams for a killer shot. This second charge is smaller to prevent the harpoon cutting clean through the catch. At the tip of a harpoon is a cast-iron fragmentation grenade with a time-delay fuse. The grenade is set to go off inside the whale for a quick, clean kill. Each steel harpoon can be re-used four to seven times. After each time it is fired and hit its mark, the harpoon is sent to the blacksmith's shop to be straightened. Eventually a harpoon, with its flukes removed, would be re-used as a killer shot.

Mick checks and re-checks his equipment. Picking up the trail of a sperm whale is an adrenalin rush. He feels the excitement in the air as he goes on deck. The crew are alert, their movements sharper, determined. Even the ship moves differently. There's more purpose to the path it carves through the swell, pushed by its 1800 horsepower, oil-fired steam engine.

As they close in, Mick marks the distance and direction of the whale on a wall chart. The ship is the centre of the map, the whale a dot in a circle. He moves the marker as Tom calls the distances.

'300 metres and close to the surface …

'200 metres …

'150 metres …

'130 metres …'

The wind-burnt face of Gordon Cruickshank, the ship's master and gunner, is fixed on the surrounding water. He takes in the information from Tom and gives instructions to the helmsman, who turns a large wheel connected to an oversized rudder, the nautical equivalent of power steering.

A pod must be approached quietly. Rush in and the whales get excited, especially the big bulls, and these are the ones who deliver the most oil. Bonuses are paid on the number of barrels of sperm whale oil produced, much the same as it was when whaling was about longboats and hand harpoons.

Males are the first to react. They flip their tails up, go straight down and are five to six kilometres away before you know it. The cows, the bull's harem, are more placid. They stick around for a bit longer before taking off.

A sperm whale can keep up a ten to twelve knot pace for a long time, before diving deep for thirty minutes or more. A high flick of the tail usually means the whale is heading for the depths; a bend in the tail, a short dive of twenty metres.

Gordon is keenly aware of the rules and regulations. It seems to him that every man and his dog want to poke their nose into his work. He faces hours of paperwork if he bags an undersized whale. He has to make excuses, in writing, telling why he's stuffed up. Then there's the quota set by the International Whaling Commission on how many males and how many females are allowed to be caught. More bulls than cows can be taken because there are always eager young bulls following the harem, waiting their chance. A pod of two hundred or more females can be governed by one large bull. Sometimes a pod like this will shrink when a young male makes his move and takes off with forty to fifty females.

Gordon can tell the difference between the sexes in an instant. A female has a pointed head, more of a taper, while a bull is squarer and bigger.

The final decision to take a whale is Gordon's. He must judge size and sex and assess the odds. Should he take this one first, or will that scare the others off? In a big swell, a whale might be in one trough and the chaser in another, with the crew getting only glimpses of what they're chasing.

At fifty-one, Gordon has spent most of his life on ships. As a boy in Aberdeen, Scotland, he read tales of adventure at sea and always wanted to join the navy or go whaling He managed to do both, despite his family's attempts to stop him. His grandfather and great-uncle had drowned at sea but Gordon was determined to follow his heart. He saw service as a Royal Navy gunner in the Second World War, then went trawler fishing off Iceland, ending up in a deep-sea trawling venture in Australia in 1949. After getting his skipper's ticket in 1968, he worked as a relieving skipper with the Cheynes Beach Whaling Company, becoming master and gunner of the *Cheynes III* a few years later. His war experience as a gunner came in handy. He could service a harpoon with his eyes closed.

When the weather is good and the chase is on, Gordon will stroll down the catwalk connecting the bridge to the harpoon deck at the bow, and wait as they close in. But it isn't uncommon for a wave to wash over the harpoon deck, so in bad weather he'll leave it to the last moment.

Based on information from the sonar, he will keep the chaser one hundred metres behind the whale, putting on speed when the whale is around fifty metres from the surface.

The whole crew feels the excitement. Everyone is on duty. They know the money is there.

Gordon likes to have two crew in the barrel — the crow's nest or lookout. Two sets of eyes are better than one.

A shout from the barrel. 'She's on the surface!'

Mick steps in to take charge of the bridge as Gordon runs along the catwalk to the harpoon. Mick tells the engine room the speed he wants, based on hand signals and shouts from Gordon. He also keeps half an eye out for the position of the *Cheynes II* and *Cheynes IV*. His skipper is concentrating on running down the whale, not looking out for the other ships. A collision would be embarrassing.

Gordon grabs the 90mm cannon mounted to the deck and kicks the locking mechanism to the off position. He quickly sights along the harpoon angled out front of the bow, his grey overalled back to the bridge amidships. Beneath the harpoon there is fifty

Gordon Cruickshank, master of the *Cheynes III*, shouts instructions to the helmsman.
Photograph by Ed Smidt.

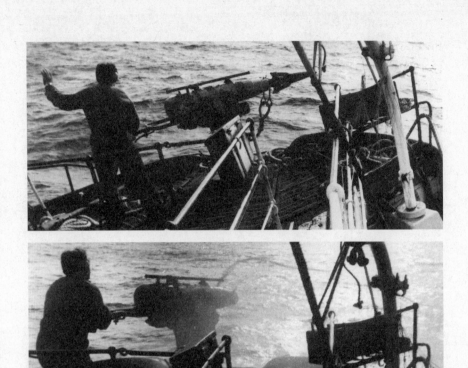

AIM — FIRE! — HIT.
Photographs by Ed Smidt.

metres of rope, the forerunner. Another thousand metres of heavier line waits in the rope locker below and will roll out when the harpoon fires. If the whale takes off, the forerunner runs through a series of springs to take tension off the rope and stop it snapping.

Muttering under his breath, Gordon shouts and signals instructions to the helmsman. The ship pushes through the waves, closing on the whale.

More frantic hand movements from Gordon. The helmsman changes course again.

The illusion of a black rock is replaced by the shiny wet reality of a sperm whale.

One more course adjustment.

Gordon pulls the trigger. A sharp explosion. The fifty-five kilogram harpoon, with its cast-iron grenade head and metal flukes, flies out of the cannon and finds its mark.

Three seconds. *One — two — three.*

The grenade head explodes and the sea turns red.

The crew always hope for this — a clean shot, a quick kill.

A winch brings the whale alongside. A spear connected to a compressor pumps the whale with air to keep it afloat. The tail is trimmed and a chain is attached. A radio beacon and flag mark this whale as belonging to the *Cheynes III*.

The chaser heads out again, butting the chop, on the tail of another whale.

Reloading the harpoon gun, *Cheynes III*.
Photograph by Ed Smidt.

The Last Whale • 19

A sperm whale alongside the *Cheynes III*.
Photograph by Ed Smidt.

2 : The Phantom : 20–24 June 1977

The Frenchman was a fifth columnist, a ring-in, they said. He came from nowhere and infiltrated the ranks of the whaling industry. On the first day of the International Whaling Commission meeting in Australia, Jean-Paul Fortom-Gouin appeared among the representatives of whaling nations carrying a life-sized sculpture of a whale's brain above his head. An activist in a business suit, he dazzled the anti-whaling movement and annoyed the world club that was the International Whaling Commission. Meetings of the commission had been held in the United Kingdom for some years. The anti-whaling lobby suspected the meeting was shifted to the backwater Australia because delegates had been given such a hard time in the UK the previous year.

Neither side knew much about Jean-Paul. He was thirty-five years old, born in Morocco to French parents, and lived in America. He arrived in Canberra with official accreditation as the representative of Panama, a country not involved in whaling at all. How the

Frenchman had persuaded the Panamanian Government to allow him to represent them at this multilateral gathering was the subject of much speculation among official delegates. The commission had been formed to look after the interests of the whaling nations. Clearly, this Jean-Paul knew nothing about the world of whaling and had no business being in Canberra.

Asked directly how he managed to get himself appointed as an official commissioner, Jean-Paul smiled and said he had business interests in Panama.

And the Frenchman appeared to have money, a lot of it. Whether the money was from real estate deals in Panama, as he once said, or from something else, no-one was sure. To the other whale defenders, Jean-Paul was a powerful figure. He set up headquarters in the best room in the best hotel Canberra had to offer and looked like the Establishment but talked like an activist.

He was small in stature but carried an air of authority. A benevolent Napoleon figure, a commander, the ideas guy who gave orders so his dreams would form substance. He would stride directly into a meeting and take control as if he was born to it. Thanks to his delegate status, he couldn't be stopped from speaking at official whaling commission sessions. Jean-Paul, in his thick French accent, invited the representatives of whaling nations to place their names in the history books by stopping whaling. The alternative was to disappear into the quicksand of time, never to be remembered.

Jean-Paul had a spiritual, almost mystical, relationship with the whales, especially the toothed mammals of the sea. He called sperm whales Cachalot, My Cachalot. He couldn't bring himself to use the term sperm whale because it was insulting to those warm blooded swimmers who had larger and more complex brains than humans. The old Yankee whalers called the Cachalot sperm whales because the spermaceti oil in the huge head looked like sperm.

Other fellow advocates for the whales recalled a story, more legend than fact, about Jean-Paul in the Caribbean when a whale surfaced beside his boat. The whale rolled on its side and peered at Jean-Paul. Whatever he saw in those eyes, it convinced him he had to stop these creatures being killed by man. After that moment, his life choices were simple: he could either keep his money to himself or he could use the money for good.

Jean-Paul wanted to speak to the cetaceans, the whales and the dolphins. He bought a dolphinarium in Florida, Santini's Porpoise School, famous as the home of Mitzi the dolphin, star of the 1960s television series, *Flipper*. He closed the facility to the public and transformed it into the Institute for Delphinid Research, creating a program aimed at understanding the whistling language of his playboys of the oceans. Jean-Paul became the Director for the Institute for Delphinid Research, a handle which would help him in the fight to save the whales.

Jean-Paul Fortom-Gouin: the Phantom.
Photograph by Aline Charney Barber.

Jean-Paul publicly spoke of a new era of cooperation, not exploitation; a new relationship between humans, the most highly evolved form of life on land, and cetaceans, the most highly evolved in the ocean. The whales had taken a different evolutionary route but, like man, their brains were the most sophisticated within their environment. The Frenchman believed the people of the land and the people of the sea would share the wealth and marvels of the blue planet.

While the anti-whaling activists and representatives of whaling nations speculated on the origins and motivations of Jean-Paul, it was a book which brought him to the whales, *Dauphin, mon cousin* (Dolphin, my cousin) by Belgian writer Robert Stenuit who made the point that many species of cetaceans have complex brains. The mystery was that no-one knew what the whales used this brain for. Jean-Paul had a prejudice for intelligence. He was interested in the smarter animals: the cetaceans, the elephants and the humanoid primates. Reading other books, he discovered that commercial whaling was smashing one species of whales after another to near extinction: the blue whale, the right whale, the humpback. Jean-Paul decided to do something about it. Stopping commercial whaling became his personal responsibility.

In 1976 he got himself invited to a meeting of marine biologists and cetacean specialists for the ICUN (International Union for the Conservation of Nature and Natural Resources), a group for the conservation of nature, at Bergin in Norway. There he made contact with members of the scientific committee of the International Whaling Commission and he was asked to write a scientific paper on cetacean neural anatomy. During his research, Jean-Paul read works by leading cetacean neuroanatomists which documented how the large toothed whales, the Odontoceti, not only had the largest brains on Earth but also the most complex by any measure: density of neurones, density of synapses, layering of neurones.

Jean-Paul wasn't a scientist. He had studied at HEC, a business school in France, and when he graduated he followed his boyhood passion of sailing by establishing a marina in the British Virgin Islands and the first bareboat charter company, Neptune Cruises. He moved to the Bahamas, a tax haven, and managed money for other people.

He noticed that many marine biologists had a beard, so he grew one too. He thought this would help him, a businessman, blend in. His lack of science qualifications didn't worry him. He got his hands on every paper written on the subject and put it together into a cohesive treatise.

No-one asked Jean-Paul about his qualifications, or lack of them, and the paper was well received at the scientific committee meeting held at Cronulla in Sydney's southern beaches before the main whaling commission gathering in Canberra. The committee made recommendations using statistical modeling to determine how many whales there were in the ocean, and how many could be caught without depleting stocks. Catch limits, or quotas, would be determined and put to the main whaling commission meeting.

Jean-Paul had gained access to the whaling commission's science and calculations used to determine whale catch quotas. He studied and dissected the calculations, concentrating on Catch Per Unit of Effort (CPUE) a method used to assess fish populations. If one year a fishing fleet spent one hour at sea and caught a hundred fish but only fifty fish the following year, that meant the fish population had fallen fifty per cent. With whaling the technology had been improving with the introduction of sonar and larger, faster boats.

Loaded harpoon with dead whale in water.
Photograph by Ed Smidt.

The tail of the whale, a sperm whale.
Photograph by Ed Smidt.

With better technology the calculation didn't work. Essentially the whalers got the same catch for more efficient effort. To Jean-Paul this meant that the whale populations were not stable. They were, in fact, declining.

With time to kill before the meeting in Canberra, Jean-Paul worked the phones trying to find out what was going to happen. He realised the real work happened before the meeting. The votes at the commission were a formality.

Jean-Paul found out Panama wasn't sending a representative to the whaling commission that year. He got hold of the Ambassador for Panama.

'You are lucky because you have a guy here, me, who is on top of the subject,' Jean-Paul said. 'I have presented a paper to the scientific committee and I happen to have close ties to Panama where I am a landlord in the capital. I propose myself.'

The Ambassador said to come in and see him. Jean-Paul found him to be pro-whale in his outlook. The Ambassador would send a cable to the Ministry of Foreign Affairs in Panama with a proposal.

One week later they said yes. Jean-Paul became the official commissioner for Panama to the International Whaling Commission.

Australian Jonny Lewis, photographer and anti-whaling activist, was drawn to Jean-Paul. This guy was scoring points and fighting the whalers at their own meetings. The Frenchman reminded Jonny of The Phantom, a mysterious comic book hero who appeared out of nowhere to fight evil. Jonny cut out a famous Phantom image where the crime fighter stands riding two dolphins and presented it to Jean-Paul in an impressive frame.

Jean-Paul spoke privately to Jonny. 'How would you like to work for The Phantom?'

'What have you in mind?'

Jean-Paul smiled. 'We're going to close Australia's whaling station.'

The idea was to start a domino effect. Australia had the last whaling station in the English-speaking world and the Cheynes Beach Whaling Company would be the first tile to fall. Other countries would follow in a chain effect until the world was free from whaling fleets. Of all the whaling nations, Australia was the ripest to stop, the weak part of the whaling coalition. Australia was more accessible than the Russians, who used factory ships far out to sea. Cheynes Beach worked from a shore-based station. The ships returned to land each day so their catch could be processed. Fewer resources would be needed to take on the Australians.

'We're going to do a Greenpeace-type operation on Cheynes and we are going to create a local group, an Australian group,' Jean-Paul told Jonny. 'We're going to bring Greenpeace in to give it an international feel. The media will think it's some sort of big international coalition. If they think we're only a dozen hippies in Sydney they won't take us seriously, right?'

The whaling company, based at the town of Albany in Western Australia's south, was working to a 1977 season quota of 624 sperm whales (508 males and 116 females). At the previous year's whaling commission meeting, the quota to take sperm whales from Division 5, the hunting area south of Western Australia, had been cut by forty-five per cent following a reassessment of the size of the remaining sperm whale population. Jean-Paul wanted the catch quota set at zero.

The Frenchman was so positive, so sure of his goals, and he had an energy around him that made Jonny feel anything was possible. Jonny put on a Save the Whale exhibition of photographs in Sydney before the start of the International Whaling Commission meeting. At his hotel room in Canberra, he showed Jean-Paul a five-minute film on dolphins he'd shot on the New South Wales north coast.

The Phantom told Jonny to use the media to crank up the pressure on the whaling company. 'I'll return within two months,' he said, handing Jonny a cheque and a pile of cash. It was more money than Jonny had ever seen, around five thousand dollars. Nervously he put it in his pocket and counted the minutes until he could get to a bank.

Jonny Lewis had been running with the more community-based groups, including Project Jonah, to fight whaling. But for Jonny, it was mostly talking, fundraising, letter writing, lobbying, placard waving. He was more comfortable smoking dope with like-minded activists than with sinking beers in the pub while planning community campaigns. The whales were gaining respectable support from middle Australia but on the other side of the continent, far out to sea where there were no witnesses, whales were still being blasted by explosive head harpoons. Jonny wanted action, not debate. The fight had to be taken to the front door of the Cheynes Beach Whaling Company.

Jonny wanted to smash through the slowly melting iceberg of public opinion. He started the Whale and Dolphin Coalition, more hippie than middle class, less than respectable. They didn't care who they upset. They were there to get the job done. Creating pressure and applying it was the name of the game. No discreet lobbying for this lot. Why bother knocking? Kick the door down and demand attention. They were edgy, right out there, and anything was possible. Focus was the key.

As a boy, Jonny Lewis had little focus. His education at the prestigious King's School in Sydney was a disaster. He was miserable and he bombed every subject. Jonny was born in the USA. His mother, an American Red Cross nurse, was on leave in Sydney during the Second World War when she met Australian Tom Lewis. After Jonny was born, the family moved from Washington and established a chicken farm at Penrith, near Sydney. His parents had split up by the time Jonny finished school and he lived with his mother in an apartment in cosmopolitan Potts Point. His father, Tom, was a conservative premier of New South Wales from 1974 to 1976 and founder of the National Parks and Wildlife Services of New South Wales.

Meant to be studying for his final exams, the second time, Jonny was mostly surfing or exploring the seedier side of Sydney, inner city Kings Cross.

Jonny found his focus in art. He was drawn to the bohemian Yellow House near Kings Cross. Started by Sydney artist Martin Sharp, The Yellow House was inspired by Van Gogh who wanted to start a community of artists in the sunshine of the south. Australia was a long way south from France and it had a lot of sunshine. It was here that Jonny took his first photographs and started his real education. He was free and happy. Now aged twenty-seven, he found his passion with the whales and dolphins, and with the Phantom, he now had the funds to give form to that passion.

3 : Willie

Richard Jones, skilled copywriter, champion of animal rights and mail-order business proprietor, was horrified when he saw a leaflet from the Save the Whale campaign. Project Jonah obviously needed help with advertising and promotion. He drove to his first Project Jonah meeting in Sydney, determined to make a full-size plastic blow-up sperm whale. He felt that would be a great image for the campaign.

While waiting for the meeting to start, Richard struck up a conversation with Ken Beatty, the man sitting next to him.

'It's my first meeting,' Richard said.

'Me too.'

'So, what do you do?'

'Well, I make plastic blow-ups.'

Richard often felt a synchronicity, a pattern in events which made connections where there shouldn't be any, when working on the anti-whaling campaign. The coincidences were too frequent to be temporary or accidental. It was as if those working on the campaign were being communicated to by the whales. There appeared to be some intelligent design at work. Perhaps the whales themselves were organising events!

There was discussion over what form the blow-up should take, a cartoon character or something more realistic. Richard got his way because he was the one writing the thousand-dollar cheque. The whale, a realistic full-sized sperm whale, was christened Willie by a Scot after the Scottish cartoon character, Oor Wullie. His first outing was on Sydney Harbour, towed by surf lifesaving boats rowed from the site of an old whaling station at Mosman Bay to the Sydney Opera House. Surf lifesaving boats were modelled on original whaling boats used in Sydney Harbour. Willie had SAVE THE WHALES written in English, Russian and Japanese in huge lettering along his sides.

Richard Jones designed an advertisement for Project Jonah, a large sad whale with the words HELP ME screaming across the top in anguished lettering. Again, Richard used his own money, this time to place advertisements in Sydney and Melbourne newspapers. The response was swift. Project Jonah membership grew and donations flowed. Richard got his money refunded and the surplus went to Project Jonah.

Willie the Whale travelled the country, floating down the Torrens in Adelaide, the Yarra in Melbourne, and the Brisbane River, focussing public support for the Save the Whale Campaign. During the International Whaling Commission meeting at the Lakeside Hotel, Canberra's best, the forty-foot whale was launched on Lake Burley Griffin, a man-made lake in the centre of the capital.

Hundreds gathered to see Willie. As the commissioners arrived for the nineteenth meeting of the commission in June 1977, they were greeted by demonstrators singing and dancing. Willie led the pack from his position on the grass outside the hotel.

Joan McIntyre, Project Jonah's American founder, pounded an

Jonny Lewis, Jean-Paul Fortom-Gouin and Richard Jones.
Photograph by Aline Charney Barber.

African drum. Later, Joan managed to get accreditation to attend the whaling commission meeting. She spoke to the final session of the meeting and played tapes of humpback whales singing. *Eco*, a newspaper started in 1972 in Stockholm for the first UN convention on the human environment, was resurrected for the Canberra whaling commission meeting by Friends of the Earth, Greenpeace and Project Jonah.

Another American, Pat Farrington, created a figure of a whale using people as her artist's brush. The whale, made up of chanting demonstrators, could clearly be seen from the upper floors of the hotel, the domain of delegates to the whaling commission.

Pat was a student leader in California in the 1960s. While campuses across America were dealing with sit-ins, demonstrations, barricades and walk outs, Pat was working with students and faculty to foster cooperation. Pat had arrived in Australia in 1975 at the invitation of the Victorian Government to help launch the fitness campaign *Life. Be In It*. She was the founder of the New Games Foundation in California based on the idea that games don't need to about combat. There are no losers in the New Games. People are still individuals but they contribute to the whole, a cooperative approach.

The cooperative lifestyle of the dolphins, where everyone plays and everyone wins, inspired her New Games. She felt a special bond, a connection, with the whales and dolphins.

During a visit to the Steinhart Aquarium in San Francisco in 1973, Pat saw dolphins, grey and white flashes through an observation window. Her first thought was, how sad. These creatures were caged, separated from the sea and their families. One dolphin went to the surface and returned with a piece of orange yarn. The dolphin dived, glided and twisted with the yarn making patterns through the water. A second dolphin joined the game, then a third.

Pat was hired two months later by the producers of the *Whole Earth Catalog*, the paper tome listing tools for inspiration, to organise the first New Games Tournament in October, 1973. At that time Greenpeace and its creators, such as Canadian journalist Bob Hunter, were heroes in the San Francisco Bay Area.

In Australia she hooked up with a group of people dedicated to saving the whales. The main organisation was Project Jonah.

Most Australians didn't know their country was still whaling until Project Jonah, an offshoot of the community-based Friends of the Earth that had its origins in the anti-nuclear movement, told them. Project Jonah started in Australia in 1976 when Joy Lee, an Australian who had been living in the USA, gave a talk at Sydney's Botanic Gardens asking for volunteers to push for an end to the killing of these magnificent creatures. Joy's passion motivated people from all walks of life. They walked in off the streets, responding to the publicity created by her talk, to offer their time and money. A room was hired at the NSW Environment Centre and Project Jonah Australia was born.

Regular volunteers came from all corners of society. Some were retired, some had jobs; some could put in a full day, some a few hours. A white-haired man, Bob McMillan, walked up the stairs to the Project Jonah office and handed over a cheque to the value of a colour television. A retired engineer, he said he'd make do with his old black-and-white set. The whales needed saving more than he needed colour.

He also offered his time. 'I'm going to handle the barbarity of this, the cruelty,' he said. He spent weeks analysing the way whales were killed and how long it took them to die from the time they were harpooned. As a result, he knew the average time for a whale to die, down to the last second. Bob was the verifier, the checker. When Project Jonah said something was happening, it was correct because he had checked it.

Project Jonah concentrated on making people aware of what was happening. Do you know how long it takes a whale to die? What about the behaviour of the whale? Explosive harpoons? They showed photographs of whales being killed and cut up. Project Jonah said Australians were slaughtering the great whales for money and there was no real human need being fulfilled. There were substitutes for sperm whale oil. Industry didn't need it. The images were powerful and Project Jonah was winning minds through the media. Newspaper coverage looked more like a campaign against whaling than straight reporting and Project Jonah burrowed its way into society, following links and relationships to the people who could bring about change, the politicians.

Project Jonah created a lobbying machine carefully following a policy of not becoming identified with any one political party. If praise was given to a politician in the government, Project Jonah would ensure the same praise went to an opposition politician within days. It was right down the middle. Anyone who helped got a personal thank-you letter.

Tony Gregory, a Sydney businessman, was asked to register

Project Jonah with a formal constitution. He was fascinated with whales, gentle animals, and he found it hard to remain uninvolved with the world's largest mammals; something growing to more than sixty tonnes in weight.

Tony put together a team of volunteers in Sydney to monitor radio broadcasts from Parliament in Canberra for anything about whales. Project Jonah would organise a reply to anything said by an MP on the same day. A committee member would write an answer to the question raised in Parliament, make 250 copies on letterhead, drive to Sydney Airport and run for the Canberra flight check-in. The faces of Canberra-bound passengers would be scanned for a likely courier. They rarely got knock-backs. The people they picked were happy to carry messages to help save the whales. Friends of the Earth in Canberra was called with the name of the passenger and met the plane on arrival, took delivery of the letters, folded them into pre-addressed envelopes and placed them in the mail pigeonholes of MPs at Parliament House. If all went well, a remark in Parliament could be answered in writing within three hours. A national election was likely to be called late in 1977 and making whaling an election issue was a priority for Project Jonah.

In Western Australia, the job was tougher. Environmentalists were viewed with suspicion. Jobs were the big issue. Closing industries meant job losses and the state Liberal government was pro-business. These conservationists were trying to do people out of their livelihoods.

Peter 'Bro' Brotherton was studying chemistry at the University of Western Australia in Perth. He'd joined the Australian Conservation Foundation but the organisation was mostly active in the distant eastern states. Locally, small groups had been established to campaign on specific state issues but as far as he could tell there was no broad-based community conservation group. He wrote a letter to the *West Australian*, the state's morning newspaper, on a conservation issue and got a call from a local group calling itself Friends of the Earth.

In late 1974, Bro and Friends of the Earth brought the campaign to the doors of the Cheynes Beach Whaling Company. He followed a tactic used in the fight to save the forests — he took the fight to where the industry was based. The idea was to engage those involved in the industry and their community. Bro wasn't too sure what would happen in Albany.

He unloaded an old station wagon and set up at the Albany Town Hall in York Street, the town's main drag, with photographs and displays about whales. It was an unsophisticated effort. There were no moving parts, only static displays. Fewer than one hundred people came through. There were some from the whaling company. A few were curious and wanted to find out a bit more, a smaller number were onside.

The only person who wanted an argument was a supporter. He tried to tell Bro how he should campaign. Don't soft-pedal, push harder, the old guy stormed.

Bro and Friends of the Earth returned to Albany in 1976, this time on bicycles. Six rode from Perth, taking three days to pedal the Albany Highway. The Australia Day long weekend in January was a scorcher. Three bikes pranged near the Albany Airport and one woman ended up on the bitumen, while Bro's bike needed a replacement wheel. They held a vigil outside the whaling station at midnight with another twenty unbruised people who arrived by car from Perth. The bent bike wheel was on display at the environment centre in Perth for a number of years in memory of that protest ride.

Friends of the Earth continued its community-based campaign. One arm of that effort, educating school children, was criticised by the whaling industry as brainwashing. Anti-whaling propaganda had no place in schools, they argued. Schools were places of learning and lefty environmental groups should stay away.

One state government minister said environmentalists should

be shot. The premier, Sir Charles Court, questioned the source of funding accessed by the environmentalists. Bro outlined the budget for Friends of the Earth. Bro said: 'If we're running on an annual budget of ten thousand dollars, the Kremlin's bloody ungenerous.'

A Melbourne journalist called Bro, saying he'd heard that Friends of the Earth was planning to attend the annual general meeting of the Cheynes Beach Whaling Company in Albany. Bro had no idea what the reporter was talking about as the WA branch had no such plans. Someone at Friends of the Earth in Victoria had made the announcement without first checking with the Western Australian members. The local branch members weren't too impressed, feeling cornered. If they didn't go ahead, it would look like they had piked, backed down. They didn't want to be dictated to from across the continent, forced into action by people who knew little about local issues and politics, so they pushed for national control of the whaling issue and Bro soon found himself coordinator of Project Jonah.

Project Jonah attracted new membership from people concerned about whales. They could join without the distraction of other conservation issues. They didn't need to worry that what they were doing may have an impact on jobs. Support grew in Western Australia. It was slower than in the rest of Australia but the single-issue organisation worked.

One of the protesters came to Richard Jones after Willie had entertained the Canberra crowd protesting at the International Whaling Commission meeting.

'Richard, can we borrow Willie?'

'What for?'

'We want to blow it up outside the corridor of the Japanese IWC delegates' hotel rooms.'

'Oh, God. It's got nothing to do with me. Okay?'

Willie was deflated using a vacuum cleaner, transported into the Lakeside Hotel via the basement lift and stored in a guest room hired

Willie the full-sized blow up sperm whale in Canberra, June 1977.
Photograph by Jonny Lewis.

by a sympathiser. Four protesters sat in the room staring at Willie and discussing the state of the world.

At 4am on 23 June 1977, the protesters put Willie in the lift and took him to the eighth floor. They reversed the vacuum cleaner and started pumping outside room 809 where the leader of the Japanese delegation was said to be staying.

They felt sure someone would wake from the whine of the vacuum cleaner and the noise of Willie popping into shape. One protester used the stairs to return to the room and call the *Canberra Times* newspaper and the Australian Broadcasting Commission to tell them what was happening. A fantastic photo opportunity, he told them. He didn't think they believed him.

At 5.45am, the lift doors on the eighth floor opened. A hotel employee stared out.

'Good morning,' a protester said.

The employee stared at Willie expanding into the corners, blocking the corridor. He mumbled a good morning as he pressed the lift button and the doors closed.

In the lobby, the employee came out of the lifts shouting. 'Big balloon on the eighth floor.'

On the eighth floor, the lift indicators showed movement.

'Good luck, Willie,' the protesters said as they took to the stairs. The four protesters hadn't slept for nearly 24 hours.

Willie ended his life at the point of a knife. Hotel staff slashed their way through the plastic to free the Japanese delegation.

A noisy funeral was held for the deflated Willie.

4 : The Scroungers

Whaling had been important to Albany since it was founded in 1826. When the first settlers arrived, sealers and whalers of many nationalities were already working the area and the whaling ships would call into the town. Local whaling ventures began within the first ten years of settlement but they were small and short-lived.

There were other later efforts, most notably by the Norwegians, but local whaling became dormant, waiting for a new cast of characters to assemble with the necessary skills, interest and knowledge.

It took awhile.

Fred Edmunds travelled with a few other blokes, camping out, pooling what little they had between them, looking after each other. It was 1930 and the depth of the Depression. They had a better chance of getting by if they worked together. The six men set up a camp near the whaling station at Point Cloates in the north-west of

Western Australia. They did some fishing, and sometimes one of the blokes would get work and bring in a few bob, enough to buy a little tucker.

Fred was in his mid-twenties. He was thankful when he woke each morning and saw the sunrise.

Sometimes the men would help the caretaker at the mothballed whaling station nearby. No-one had the money to buy whale products during the Depression, so the owners had shut down the boilers and equipment, hoping to restart operations when the markets recovered. The equipment needed protection from salt and spray. Chipping away the rust and repainting was a never-ending task.

When times were better, Fred wandered Western Australia's eastern goldfields and got a job working a steam plant powering the mines. He started as a fireman, sweat mixing with coal dust, as he fed the fires to heat the boilers. He kept his head down and worked hard. The bosses kept him on and eventually he qualified for a third-class steam ticket. He worked harder and got a second-class ticket. Finally he gained an unrestricted first-class steam ticket, along with a good working knowledge of engineering and making do.

Charlie Westerberg grew up in Albany and left school in 1933, aged fourteen. The eldest of three boys and one girl, he was luckier than a few of his schoolmates. Some fathers, seeing no way to feed all their children, told the oldest that they couldn't keep them any longer. The fourteen-year-olds were turned out of the house, a few shillings in their pockets and a couple of blankets to keep them warm at night. They lived on the road with the long lines of men and boys looking for work.

Charlie was called Snapper because the old chap was also a Charlie, and they were a fishing family. The second son, Gordon, was called Gardy (short for garfish); and young Neil was, for some reason no-one remembered, called Bill.

All the local children knew each other, and many were related. Snapper's family was close to the Birss family, who lived at Emu Point. They had a son, Ron, two years younger than Snapper, and both families went fishing. It was an existence; there was no such thing as a rich fisherman.

During low tides, all the kids would play cricket on the hard sand at Emu Point. When the tide rose, they sought adventure in the nearby bush or went swimming. You learned by the sink-or-swim method. In summer, the locals were joined by the post-harvest exodus of farming families escaping the dry heat of the inland for the cool breezes of the coast.

Snapper and his father, Charlie, fished the coast, catching whiting, leatherjacket, garfish and skipjack. High seas and submerged rocks weren't the only hazards they faced. During the winter months there were so many humpback whales in King George Sound they had a job dodging them in the gloom of early morning. Snapper and his father often thought about those whales.

After the Second World War, the salmon fishing industry got going in Albany. A cannery was started and the Westerberg and Birss families were doing well fishing off Cheynes Beach. The money started to come in and they bought a tugboat that had been built for the war but never used.

Ron Birss and his father, George, used the tug to trap for leatherjackets outside Breaksea Island at the mouth of King George Sound. They saw a lot of humpback whales in the three months of winter. They watched with great interest when a group started the Albany Whaling Company in 1947.

Ches Stubbs grew up in what was called Buckland Hill, later renamed Mosman Park, a suburb of Perth between the Swan River and the ocean. He was always down at the local jetty fishing in the river, and after the Second World War he went into commercial fishing. With more keenness than experience, he fell into the job as a gunner on a whaleboat, the *Wadjemup*, and when the Albany Whaling Company brought the boat down to Albany he came with it.

The prospect of a whaling industry being established was big news. Ches and his friends hunted humpback whale close to shore in King George Sound and their exploits were trumpeted in the *Albany Advertiser* on 24 July 1947: WHALE CHASER WADJEMUP MAKES FIRST KILL: FORTY FOOT HUMPBACK TAKEN NEAR ALBANY.

The newspapers portrayed the event as a classic battle between man and beast. It had started at 4.10 pm and lasted eighteen hours. Ches got in the first harpoon. It took another three harpoons before the humpback was finally caught, ready to be dragged back to land by the underpowered ex-navy plywood boat, the *Wadjemup*. To keep the whale afloat, Ches climbed onto the carcass and cut a hole to take a hose connected to an air compressor. The air escaped through the harpoon wounds as fast as they pumped it in. Ches had to clamber back onto the humpback, plug the holes with rags and scramble to safety.

Back on land, it took six days to flense that first whale. They had no heavy gear and had to strip the blubber by hand, before cutting it into little pieces.

The under-equipped Albany Whaling Company was turning belly up. They had a 50mm gun from Norway designed to catch twenty-five foot whales, but the Australian Government's rules stated that the catch should be a minimum of thirty-five foot in length. They took six whales in two years, hardly a commercial success, but those involved in the first whale-killing by an Australian crew in Australian waters for many years were, according to the newspapers at the time, heroes, or at least, enterprising people.

Albany fishermen held a lot of meetings after the Second World War in an attempt to get people to band together to get whaling going again. The Birss and Westerbergs stuck at it more keenly than most.

A friend of Snapper's, Norwegian Larry Johnson, said he'd write to a friend in Norway who had been whaling in Albany before the war.

'Tell your young friend to get off his backside,' came the reply from the other side of the world. A Norwegian whaling operation was starting in the north-west of Western Australia. 'Get up there and see how it's done from the catching side. The shore side will sort itself out.'

Snapper got a job as a deckhand at the Nor' West Whaling Company at Carnarvon for the 1949 season. His workmates were as green as he was. The Australians didn't have a clue. The Norwegian experts were happy to explain the why and the how, and so Snapper was able to pick up the tricks of the trade. Bringing a harpooned whale alongside could be tricky. A whale with a the harpoon sticking out of its side should be towed to shore with the harpoon pointing away from the vessel. Do it the other way and the harpoon would grind a hole in the hull.

Each day was different. The behaviour of the whales fascinated him. One day a whale came up under the stern and broke one of the propeller shafts. Snapper thought the ship was going to shake to bits.

He saw how protective the female humpbacks could be, sometimes carrying their young on their backs out of the water. Once he saw a calf on its mother's head and she was going full speed — the red butt where sharks had bitten off the calf's tail clearly visible.

The humpbacks were so thick the whalers could pick and choose which to kill, looking for the biggest and fattest, but they wouldn't take a whale with a calf.

When Snapper spoke to the Norwegians about his family's idea of whaling in Albany, they told him he should be able to buy a vessel at a good price. Most of the whaling nations in the northern hemisphere had their eye on the Antarctic, and were moving to larger ships for the long voyage south. Many of their shore-based vessels were redundant.

Fred Edmunds, having worked a variety of jobs after the Second World War, found himself helping to rebuild the whaling station at Point Cloates where he and his mates had camped during the Depression. In 1949 he met a young bloke at the Nor' West Whaling Company. The whalers called him Westy but he said his family called him Snapper.

Snapper told Fred about Cheynes Beach, forty miles or so east of Albany, where his family and the Birss family had cottages and caught salmon. Storms would often gouge out parts of the beach, he said, exposing mounds of whalebones. Along the coast they found abandoned tripots — large cast-iron pots used to boil whale blubber — that had been left behind by nineteenth century whalers. If the area had been used for whaling before, it could be again.

While Snapper was up north learning to catch whales, the rest of the family was busy with the shore side of things. Using a truck and a four-wheel drive to cart sand, shovelling it out by hand, they built an area for a flensing deck It was slow, hard work. They registered a business name, the Cheynes Beach Whaling Company, and sold the tugboat to the government to help finance the move into whaling.

Fred Edmunds knew what was needed to make a success of whale processing. First you needed winches strong enough to pull a sixty to seventy tonne whale up a slipway to be flensed and processed. For that, a solid foundation was essential. When he saw where the Albany families planned to put the whaling station, he knew it wouldn't work. Cheynes Beach was on soft ground. It would be impossible to put in concrete to hold winches big enough to pull a whale up a steep slope.

The two families thought about it for a while. They took Fred to Frenchman Bay where there was a flat rock sloping down to the water. The natural rock formation was a marvellous anchor for machinery and would save on concrete and reduce costs. The company didn't have a lot of capital to play with.

Fred, a bush designer, self-taught engineer, foreman and a jack-of-all-things-construction, became works manager. He brought in Syd Reilly, an accountant from Perth. Wally Saleeba, who had the Ford motor vehicle franchise in the state, also came on board. They used Fordson tractors, the world's first mass-produced tractors, to prepare the ground to build the whaling station. Many potential investors were frightened off because it was an Australian venture. The Norwegians were the ones who knew about whaling. What would these Australians know? It made it difficult to bring in capital but a group of Albany locals joined as investors, including Dr Harry Hanrahan, dentist Rupert Holmes and Dan Hunt at the cannery.

Fred Edmunds put in some of his own money. The company couldn't get credit because they didn't have income, so they opened a bank account and worked with cash.

The station grew organically, each piece conceived in the mind of Fred Edmunds. There were no plans to follow, no drawings to consult and no structural engineering rules. Fred worked out what had to be done and got on with it. After the Second World War there was little around in terms of materials, but plenty of equipment was lying idle if you knew where to look. And Fred did.

'It was my design because I knew what we wanted,' Fred says. 'I knew what had to be done. It was a matter of getting everybody to see likewise and work towards putting it together. Everything had to be made. We had to know enough about how to do it. The deck, the anchorages for the winches, all the steam plant, all the digesters, all had to be made or improvised from whatever you could get. And it worked.'

He had no formal engineering training but had absorbed knowledge from the old blokes in the goldfields and the time he helped rebuild the whaling station in the north-west. His steam engine knowledge was invaluable because town water and power had not yet reached Frenchman Bay. They used steam to power the whole operation, drawing water from a soak at a nearby swamp.

A master scrounger, Fred put each piece of the puzzle together from a different source: farm machinery, odds and ends, Second World War surplus, second-hand pipes and brass fittings. They cut timber from the hills to the west of Albany, around Torbay, for structural work, the flensing decks and the stumps to hold it all up. From the mining town of Collie, north-west of Albany, they brought metal tanks from a wheat distillery built during the war, plus a truckload of valves, pipes and other fittings. From Ora Banda in the goldfields came Lancashire boilers, a steel mast and lots of steel cable. They got hold of army surplus Nissan huts — corrugated tin structures — and slept in those, working up to midnight to get the job done. The project took twelve months.

The Westerberg and Birss families wanted Johannes Andresen, Snapper's Norwegian friend, to go to Norway to get them a proper whale-chasing boat.

'I'd like Westy to come with me,' Andresen told them.

Snapper and Andresen went by ship to Southampton, train to London, plane to Oslo and train to Sandefjord. They arrived on 8 October 1951. While Snapper's family in Australia was sweating from the labour of building a whaling station, Snapper's thin Australian clothes were offering little protection from the Nordic cold.

Andresen had all the contacts, and had established that there were four whale chasers for sale. Antarctic whaling was the frontier now. His father-in-law, who had returned from six months at sea, agreed to look over the *Toern*, the ship Andresen had his eye on. The verdict: 'She's sound as a bell.'

Snapper didn't do anything without Andresen's advice. The Norwegian had forgotten more than Snapper knew about whaling.

Snapper organised a cheque for fourteen thousand pounds and the *Toern* was towed across the fiord to a slipway. The scraping, painting and equipment-stowing began. They loaded ropes, whalebone saws, grenade heads for the harpoons, an oil separator, and endless bits and pieces.

The ship sat low in the water and Snapper worried about the load. He questioned the engineer about the fresh water he was pumping aboard. 'She's pretty deep in the water,' he said.

'I know,' the engineer replied. 'But this is the last good water you'll see till we get to Australia.'

They needed a crew to sail the *Toern* to Albany. Andresen put an advertisement in the local newspaper for anyone wanting to emigrate to Australia. Snapper had checked with Australian authorities before he left home and was told, 'Norwegian? No trouble. They're all right.'

Snapper thought he heard bells as the *Toern* left Sandefjord and headed out the passage through the fiord. The first mate told him they were bells driven by the power of the waves. If they got louder during a fog, the ships would turn around and go back.

When the ship was one day at sea a warning went out. A nor'easter was coming, gale force, straight from Siberia. Temperature drops to minus thirty-four were expected. All but three of the seasoned crew were seasick. They stayed on deck with the first mate, Harry Hansen, and the skipper, Rolf Elnan. The rest of the crew kept below.

Then a warning came over the radio: 'Attention. Attention. German mines have broken loose.'

The warning meant nothing. It wouldn't have mattered whether there were mines there or not. The snow was so thick, Snapper couldn't see past the bow.

The cook, before he became too ill, made a storm soup of porridge and raisins, and that was all Snapper ate for three days. He catnapped on the table in the galley or chewed sweet tobacco, and acted as second mate for the next six weeks.

They landed in Algiers to take on fuel, went to Port Said in Suez, through the canal to Port Tewfik, on to Ceylon and Jakarta, then through the Sunda Strait and down the north-west coast of Australia. They rounded Cape Leeuwin on the south-west tip of Australia on 17 May 1952 and landed in Albany the next day.

The ship was renamed *Cheynes* and started hunting humpbacks.

A whale chaser, Albany, Western Australia.
Photograph by Ed Smidt.

5 : Ahab of the Southern Ocean

Ches Stubbs joined the Cheynes Beach Whaling Company early in its operations. He was the most experienced Australian gunner after his brief stint with the Albany Whaling Company in the late forties, and he eventually became skipper of the *Cheynes III*.

Ches was known for the long shot — firing the cannon and sending a harpoon in an arc like an arrow from a bow. When he had his eye in, he would astound the crew with the accuracy of some of his shots. It was incredible magic. When the harpoon fired, everyone believed the shot couldn't possibly find its mark at that distance, but it always did.

When the sun was shining and the ship cruising, Ches was easy to get along with. He was known for his generosity and humour, but as soon as he sighted a whale — *his* whale — the air around him turned blue. If he thought a deckhand wasn't doing his job properly, he would suggest the man have more porridge for breakfast or stop staying at his girlfriend's place all night or both.

On the bridge of an Australian whale ship in rough weather.
Photograph by Ed Smidt.

'You're not worth a gallon of ant's piss,' he would say.

If Ches thought the bloke in the barrel wasn't giving the correct information on where the whales were, he would say, 'You couldn't steer a brick shit house up York Street.'

And if the lookout still didn't perform: 'If I could turn this gun around, I'd blow you out of the barrel.'

When nothing was going right, Ches would curse, 'Fuck me up a blind bullock's arse.'

He would berate those who even *thought* about being tardy. His swearing was so intense and imaginative that some people would physically cringe in the presence of it. Occasionally he would sack the lot of them, useless bastards. Later, when he calmed down, he would re-hire the whole crew. Couldn't bear the idea of them being out of work.

Ches kept a sulphur-crested cockatoo on board as a pet. Usually it was attached by a leg tie to a perch built for it on the bridge so it could be close to Ches. But sometimes Ches would undo the leg tie to give the bird a break, and it would take off, roaming the ship looking for victims.

The chaser was run along casual lines of command. The skipper was called Ches by everyone, and no-one had heard of safety boots. A lot of the deckhands wore thongs. One moment one of the men would be steering the ship, the next he'd have a searing pain in his foot. The cockatoo had struck again.

Once, without warning, a whale surfaced right in front of the ship and Ches raced down to the gun deck and shot it. The cockatoo, hanging from one of the stay wires on the funnel, startled at the sudden explosion and ended up in the water. The crew shouted to Ches, but there was a whale on the line and work to do.

One crewman wanted to dive in and rescue the bird. The others talked him out of it and they used a boathook to drag the bird in by the neck. A deckhand tried to give it beak-to-mouth resuscitation but the bloody thing was dead. No-one wanted to be the one to tell Ches. He was very fond of that bird and it was a while before he could see the humorous side.

On 20 October 1965, Ches, already larger than life, was about to gain the status of legend. It was getting late in the day. He didn't have his eye in, that space in his mind where he knew he had that whale when he fired the cannon. The whales were elusive and the weather was wearing from bad to worse.

A wave washed over the harpoon deck as he was taking aim, and sucked the forerunner out of its box. A kink in the line, a coil, snagged on his left leg as the harpoon shot out of the cannon. One moment he was standing, the next he was on the deck. He watched the blood gush from where his left leg used to be. What a sod of a way to die, he thought. He was bleeding to death on a ship in the middle of the Southern Ocean, at least three hours' away from expert medical care.

First on the gun deck was James Hart, an Irishman. James was universally known as Paddy because Ches couldn't for the life of him remember the young bloke's name. Paddy was shocked. Ches kept trying to stand up as if he couldn't believe he'd lost the bottom half of his leg. It took Paddy some time to calm him down, keep him sitting and help him realise he'd been hurt.

The crew applied a tourniquet to slow the bleeding, gave Ches two morphine tablets and wrapped him in blankets. Paddy called the other two chasers to let them know there'd been an accident.

In the engine room, Ches's son Mick knew something was wrong. Normally, after a harpoon had been fired, there would be instructions from the bridge via the telegraph: half speed, full ahead, something. There was nothing.

Mick shouted into the voice pipe. 'What's happening?'

No answer.

Eventually Alf Lawrence, the chief engineer, told him the bad news. The crew wouldn't let Mick up on deck to see his father. They thought it best he stay below, stay busy.

John Bell, the spotter pilot, had always said flying was mostly boredom. This day started that way. He'd seen Ches take the last shot and, low on fuel, had turned the plane for home. A few minutes later the radio crackled into life. Could he land and pick up Ches? It would take too long to get him to town by sea and he would bleed to death in the meantime.

John dropped a rubber life raft next to the chaser. The crew brought the inflatable alongside and lowered Ches into it. John got the Cessna 172 down on the water intact but was worried about take-off. At sea level he had a much better idea of the problems he faced. The aircraft, called a seaplane but more suited to the still waters of a river or estuary, was underpowered for the peaks and troughs of ocean swell. But the first problem was how to get Ches into the plane. Because of the wingspan of the Cessna, he couldn't taxi in close to the chaser.

At the whaling station, the radio message came through: 'The skipper's lost his leg.'

Andy Woonings, the works manager, heard the message and ran into station manager Jock Murray's office where the radio had two-way capability. Jock radioed John Bell.

'I'm already on the water,' John said.

Alf Lawrence went with Ches in the life raft, using bits of wood as paddles, but the flexible craft had been designed for floating, not canoeing. As hard as he tried, Alf couldn't cover the two hundred metres to the plane. The life raft went around in circles.

On the ship, Paddy could see they weren't going to make it. Although the water was full of blood from the harpooned whale, he dived in, grabbed the raft and swam it across to the plane.

'I didn't even think about sharks at the time,' Paddy says. 'It was only on reflection I realised that I might have been in a little bit of danger.'

Ches was in a fog. The painkillers were working. He had only the faintest realisation that he was being loaded aboard an aircraft. Alf held him as they closed the door on the plane. John pushed the machine as hard as he could but there wasn't enough power to get into the air. The plane wobbled. One wing dipped into a wave and a few inches of wingtip went missing. They would have to lighten the load if they were to lift off the water.

'You'll have to get out,' John told Alf over the roar of the engine.

But Alf couldn't swim. John brought the plane as close to the chaser as he could and Paddy dived in again. After towing the raft to the plane, he got Alf in and started swimming back. Before he knew it, his jeans were halfway down his legs. He kicked them off and kept going.

In the Cessna, John recalled reading an article about sea landings and take-offs during the Second World War. Warships would sail in a tight circle to flatten the water, and he had seen chasers do the same when they went after whales. He radioed the other chasers, who had arrived on the scene and were standing by to assist. Pump the bilges and spread oil on the water, he told them. They quickly did as he asked and black oil calmed the seas.

John headed the Cessna parallel to the waves into the relative calm left by the chasers. The swell smacked against the floats, sending shudders through the frame of the plane. The wing tips cut the sea as the plane tossed from side to side.

Ches looked up at the window and saw blue water. The plane was semi-submerged on one side. He shook his head and took two more morphine tablets.

A large wave loomed ahead and John took a chance. Engine straining, he turned the spotter and caught the wave like a surfer. It tossed the plane up and John pushed his machine into the air, quickly gained height and turned for Albany.

The crews on the chasers were relieved to see the plane shake off the swell, but Paddy was sure he heard the engine miss during take-off.

In the air Ches was swearing his head off, his booming voice confined in the cockpit. John closed his mind to the noise from the seat beside him and concentrated on flying.

Ches passed out halfway to Albany, quietening things down for a bit, but then fell forward onto the controls. John strained to pull the heavy man off, flying the plane with one hand and holding back the dead weight of Ches with his other. Luckily Ches woke again, freeing both John's hands for the water landing at the channel joining King George Sound to Oyster Harbour. He finally nudged the plane into the sand of Emu Point.

A handful of people, hearing the noise of the aircraft, had gathered to watch. The ambulance officers waded in to carry Ches to shore. His leg, wrapped in cloth, followed separately. Ches's bellowing voice, spitting out a stream of creative invective, orchestrated the manoeuvre, telling everyone what they should or should not be doing as Dr Alan Fitzpatrick finally got Ches into the ambulance.

Extract from the log book, Cheynes III:

Wednesday, 20 October, 1965
1210–1220: Master's left leg fouled in forerunner and left leg amputated about eight inches above ankle.
1300: Return to base under acting chief officer [Kurt Gustafsson].

'We did some damage to the aeroplane,' John said. 'We had to put a new wing on it.'

'From the time I had my leg sawn off it was fifty-five minutes before they got me to hospital,' Ches says.

Legend has it that Ches was up and about a few months later, saying he hardly missed his foot, the artificial one was as good as the original, and was looking forward to the next big dance night. In reality, he was in Albany Hospital for two months. He almost lost his nerve but forced himself back to work. When the *Cheynes III* had to go to Perth for a survey, Ches put up his hand and took the ship to Fremantle to be slipped. 'I think if I hadn't gone on that trip I would never have gone back.'

He worked for another ten years, gathering more stories of the sea. He saw a few albinos, like the elusive white whale described in *Moby Dick*, the classic novel by Herman Melville. He got close enough to one when he harpooned it to see that it wasn't exactly white, more a creamy colour.

6 : The Whaler's Life

John Bell was furious with the Russians. The Soviet fleet vacuumed anything that moved as the factory ships went past Australia's sperm whale hunting grounds. There wasn't a decent-sized whale within three hundred kilometres.

The state-run Soviet whaling fleet operated with military precision, using a team of three ships like their Australian counterparts, but that's where the similarity ended. The bigger, faster Soviet whale chasers worked with a factory ship, had no need to return to port, and easily gobbled up an enormous number of whales in a short time. In contrast, the Australians had to tow their catch to port each night, and each ship valued their independence and had individual ways of going about their business. John thought a casual observer could be forgiven for thinking the three Albany whale catchers were actually run by different companies.

'The Russians had obviously taken sperm whales all the way from Tasmania to here,' John says. 'We went for four to five weeks without seeing a whale. They were in international waters on their way to the Antarctic but they should have stayed away from Australia's operating area. We complained, of course, through the official channels, but I don't think anything ever happened about it.'

The Albany whalers speculated that it was payback for Australia voting against the Russians at an International Whaling Commission meeting. No-one could recall exactly what Australia may have done to upset the Soviets but this was the 1960s and the height of the Cold War. Everyone suspected conspiracy when it came to the Soviets.

The Russians maintained they were within their rights as they were operating in international waters, fifty to seventy kilometres offshore. At this time twelve miles (twenty kilometres) was the limit of Australian waters.

On the Russians' return trip from Antarctica it was worse. Their ships ran rings around the Australians, taking whales from in front of the Australian chasers. John watched it all from the air. Not content with just sailing past on their way home, the Russians were thumbing their noses at Australia.

The Australian whalers relied on John Bell's expertise. They had confidence in what he told them. If he said a whale was a big one, it was. If he said it was fifteen kilometres away, it was. John was steady, reliable and conservative in his approach to his job. He used plastic bags containing bunker oil — a thick, sticky black liquid — mixed with a powerful fluorescent dye to mark an area where a deep-diving whaling had gone down. Some disappeared for sixty minutes or more. The chasers would go to the area, hoping they could pick up the whale when it surfaced.

In one incident, John buzzed the Russian factory ship, got in position and let fly with the bunker oil bombs. The action was more a token protest than any real attempt to stop the Russians, but he felt better as the sticky liquid splattered the deck of the Russian whaler. His aim was good — so good that he managed to cover the captain of the Soviet whaling fleet.

Everyone thought it was a great joke until a Commonwealth policeman, who had travelled all the way from Canberra to Albany, knocked on John's door.

'The Russian Embassy in Canberra wanted me charged with an act of aggression on the high seas,' John says. 'It was a bit worrying for a while, not knowing what the penalty for this sort of thing was. It sounded a bit like I should be shot at dawn or something.'

John's colleagues told him he would probably be hanged for piracy. However, the Soviets had little influence in Canberra. John wasn't charged. The Cheynes Beach Whaling Company got a polite letter from the Department of Civil Aviation pointing out that it was against regulations to throw something out of a plane.

The White Star Hotel on Stirling Terrace, run by Molly Moyes, was closest to the Town Jetty by a few metres. The chasers would drop the catch at the whaling station and make straight for the jetty.

Molly looked out for all the single men, keeping a spare bed at the hotel in the off-season when they weren't sleeping on the chasers. She listened to 6VA, the local commercial radio station, for the chasers' estimated time of arrival. The radio would also announce whether whales had been caught and whether flensers were needed at the whaling station. If the chasers returned to port after closing time Molly left a carton of beer for the boys at the jetty. There was no danger of anyone taking off with the beer; no-one would dare.

'Molly was the whalers' grandmother,' deckhand Peter 'PJ' Johnson says. 'There was no swearing in front of her or you'd be out the bloody door. When Molly spoke, Molly spoke.' PJ would hand his cash pay — 'no bloody cheques accepted' — over the bar to Molly. Bar bills were paid from the balance in the safe, and money was rationed over the off-season. 'By the time we went whaling the next year, I'd still owe her two hundred quid. We didn't believe in banks.'

Life was a tight circle for the whalers: sea, pub, payday, pub, day

The whale chasers at the Town Jetty, Princess Royal Harbour, Albany, Western Australia. Photograph by Ed Smidt.

off, sea, pub. On land everything was within walking distance of the Town Jetty. They dropped into the Hub on York Street to buy a new shirt, or went for a haircut around the corner, but mostly they would just wear a rut between the jetty and the White Star.

'If you were single, you were allowed to stay on board as security for the ship,' PJ says. 'We had free electricity, stove and a bed. If we had a good cook, he'd save up the Carnation milk and the coffee for us to have in the morning.'

Whaling was the sort of job that brought the unexpected. Paddy Hart liked the camaraderie. They lived together, worked together and drank together. If they had a fight, they had a fight together. But being the toughest carried a certain responsibility. The shearers fancied themselves but they weren't in town a lot. The whalers were in good shape and certainly had a reputation for drinking and fighting.

'There was a bit of macho image, I think, to see who was the toughest,' Paddy says, 'the whalers or the shearers. On Friday nights, if whalers were in port, it came to a test of strength.'

One of the toughest was Keith Richardson. One night the crew hit the Town Jetty at nine o'clock and bolted for the London Hotel, the pub in vogue at that time. At the bar was an Englishman in brand-new khakis. The whalers had him pegged as a wanker as soon as they saw him. He was too neat and his mouth was bigger than his body. He was trying to pick an argument with Keith. The other whalers grinned as they watched. Some shook their heads. The Englishman kept pushing Keith to come outside.

Keith ignored the buzzing voice for a while but the Englishman was insistent. Keith didn't look at him. Apart from being strong and fit, Keith spent some of his leisure time teaching judo. Finally, he put his glass on the bar, wiped his hands, got down on the floor and did ten one-armed push-ups. He got up and turned to the Englishman. 'Still want to go outside?'

The Englishman walked straight for the door and was never seen again.

There was more activity on the jetty on payday. Wives positioned themselves at the entrance to the jetty, blocking the way to the White Star. They wanted to get their hands on the pay packet before it was lost over the bar.

One married man spotted his wife waiting for him. He borrowed a motorbike helmet from another crew member and jumped on the back of his bike. They went straight down the jetty, flat chat, and came home four days later, the pay packet missing in action.

Behind the bar at the White Star was a whale's penis. There was another at the neighbouring George Hotel. These were considered high-end souvenirs and a source of great amusement. The biggest one was said to be fourteen foot long. The flensers at the whale station would peel them, fill them with sand, put them in the heat of the boiler room to dry out and sell them for up to ten dollars per foot. Some ended up tied to tourist buses. The flensers liked to watch the buses drive away, a whale penis bouncing behind.

One bloke had a unique way of clearing the bar in winter when there was a fire burning. The charge used in a harpoon head was a bag of gunpowder weighing 650 grams and he would stuff one of these in his shirt, then come up to the pub and chuck it in the grate. There was no explosion but the black power would burn fast, sending clouds of creamy smoke into the room.

Over lunch on board the *Cheynes III*, skipper Gordon Cruickshank talked about whales, and as he talked he watched the reporter and the photographer. He could see they were fascinated by the hunt for a whale but the big question was: would they last through lunch? A meal was the danger point for seasickness. Many made it through the morning because they were outside in the fresh air, moving about, but lunch meant they had to come inside and test their stomachs. Most lost it when faced with food.

The cook glared at those in the mess room: Gordon, Mick Stubbs and the two visitors. Lunch was serious business, not for testing the mettle of new chums. The food was the same in each mess — the crew would revolt if the officers were given better food. Today's main meal was sausages, made a sickly shade of green by the addition of a packet of curry powder.

Ed Smidt, photographer with the local paper the *Albany Advertiser*, grew up in Albany. He and his school friends admired the whalers and getting a job with them was the pinnacle of many boys' ambition. The money was great and the whalers were all tough men. Ed had no trouble with lunch. He cleaned the plate.

But I was the young reporter, and I had grown up in Perth. I carefully ate my way around the slime-green sausages, picking at the mashed potatoes and vegetables. Touching the sausages, much less putting them near my mouth, would have tipped me over the edge. The trick is to keep eating. If it's going down, it's not coming up.

Ed Smidt and I had organised the trip with the *Cheynes III* months before. The Cheynes Beach Whaling Company was at first suspicious, suspecting our motives for wanting to do a feature article at that time. The buzz of the anti-whaling movement was getting louder and the whaling company had been getting more requests from Eastern States newspapers to witness the whale hunt. Our real motive was to get out of the office and away from the daily grind of news creation. We'd try to organise something each month for at least a full day away and return with a two-page story with photographs. The date of the trip, 20 June, sounded familiar when it was eventually proposed by the whaling company but it wasn't until the night before that I realised it was my birthday and that the *Cheynes III* was a dry trip. No chance of getting an alcoholic drink. Ed and I had no idea that the International Whaling Commission was meeting in Canberra at the same time.

Gordon smiled and answered questions as he ate. He too cleaned his plate. He might have the odd dig at the cook but he was careful because a good cook was the fabric of a ship. Lose the cook and the crew would unravel. The chasers were home for many of the crew and they ate better food than anything they could put together themselves. It was typical bloke food: steak, chops, sausages, bacon and eggs, roasts, perhaps fresh fish once a week, all served with potatoes and vegetables. A concession to ethnic preference was the pot of percolated coffee on the stove. The Scandinavians couldn't operate without it. The Australians kept to instant coffee.

Gordon was explaining that a clean kill is in everyone's interest. A bad shot creates a messy business, costing time and money. The chaser then has to move in for a killer shot — a wooden harpoon with an explosive head, no forerunner — to make sure the whale is dead.

And no-one wanted an undersized whale. No profit, no bonus,

and Gordon would have to spend hours on the paperwork, setting down the excuses. He would also be fined as much as five hundred dollars and everyone, from the International Whaling Commission to the Western Australian Department of Fisheries to the management of the Cheynes Beach Whaling Company, would require a detailed report on why an undersized whale had been taken.

Gordon hated making excuses. One report about an undersized whale began: 'This whale being undersized was as much a surprise to me as it was to the whale when I shot her. The whale was on its own, which is unusual for cows, and at the time it looked quite big enough. If I had thought otherwise I would not have shot it.'

The bottom line on size was thirty-four feet and six inches. Authorities allowed a six-inch error margin but anything more and the skipper was in trouble.

Inflating and buoying a dead whale could be dangerous. Mick Stubbs was always wary during this procedure. He'd woken up a couple of times with a bad headache, flat out on the deck trying to work out where he was and what had happened. An unexpected wave could bring a whale's tail down on the unwary.

Once killed, whales were towed by a rope attached to their tails, but in rough weather the tails sometimes broke off as the chasers headed home, but they couldn't stop in case the tails of the other whales they were towing broke as well. They would go out the next day to retrieve the lost whale, which was a shipping hazard.

The *Cheynes III* with its catch.
Photograph by Jonny Lewis.

Mick Stubbs recalls a time when a harpoon went straight through the whale they shot. They were bringing it alongside when he saw something strange hanging out of it. The whale was giving birth.

'The whale was dead,' Mick says, 'but it was still contracting. Next thing this calf popped out. The other cows came in and grabbed hold of it and took the calf away.'

The crew had gone quiet. Some shook their heads, others turned away. Without discussion, they took no more whales that day from that pod.

'The intelligence of those mothers was amazing,' Mick says. 'They knew the calf was going to come out. That was the worst. It put me off. All the blokes were upset. But you can't tell if a sperm whale is pregnant.'

Life for Mick was full-on. When they caught whales, the days stretched. 'You picked up the catch on the way back and dropped them off at the station. You had a few hours, turned around and did it again. It could be two or three in the morning when we got back.'

I asked Mick his view of the environmentalists who said whaling should stop.

'Those people who want to save the whale, they don't know the first thing about it. For Christ's sake, they haven't seen whales and they don't know how long it takes for a whale to die. I can honestly say that I haven't seen a decrease in the number of whales out there. When you see three hundred whales you can only put it down as approximately. There might be four hundred or there might be two hundred, but there are a bloody lot of whales going past there.'

7 : The Studio : June 1977

Aline couldn't go past the building; it had a pull like a magnet and something strange was going on inside. She couldn't see much of the interior of the cavernous warehouse on New South Head Road but what she could see was fascinating. People were moving about and there was a haze in the air. The first time she approached the white-painted brick building, she glanced to her left, but that quick look wasn't enough so she doubled back and stopped at the window. Nothing she saw made sense. Five times she tried to pass the building. Finally she made up her mind and walked in.

Her eyes took in the scene. The walls: large oil paintings; black-and-white photographs. The floor: painted outlines of whales and dolphins. The people: Aborigines with face paint, and white Australians sitting on the floor barbecuing lamb chops over an open wood fire.

'What's going on in here?' she said. 'Is this an art gallery?'

And so the dance group from Mornington Island met Aline

Charney. Aline thought she'd got a feel for Australia in the three weeks since she'd stepped off a thirty-five foot racing yacht after the eight-day crossing from Noumea, but this was something else. She had walked in on an all-night party of talking, singing, telling stories and cooking on the floor.

A tired voice from the back of the room asked, 'Are you American?'

'Yes.'

'Well, come in,' Jonny Lewis said.

Jonny had his photographic studio in the Edgecliff warehouse in Sydney's eastern suburbs. He introduced her to the dancers and to Tom Barber, an architect, who also worked and lived in the former wool store. Jonny and Tom had been helping with Bush Video, an early form of cable TV, at the Aquarius Festival in Nimbin. They had come into contact with the dance troupe through their anti-whaling campaigning with Project Jonah. The Aborigines knew about whales and dolphins, Jonny told Aline. They had a strong connection with the environment and had a lot to teach about man's connection with animals and the land. Jonny had played his short film, *Dolphin Dreamin'*, about dolphins on the north coast of New South Wales. The Mornington Islanders had loved it and sang their dolphin song to the images.

Jackson Jacob, one of the elders of the Lardil people — now cooking on the floor — told stories about his father naming his canoes Dewn, the word for dolphin. His father would sing a sacred song, a slow vibrating whistle, as he walked along the beach. The dolphins would respond by herding fish into shore and the young boys could spear as many fish as they wanted in the shallows. Jackson's father would slowly clap his hands and watch the dolphins come in closer for their share of the catch.

Aline was born in Berlin to holocaust survivors, and the family moved to the United States in 1949, where she grew up on a chicken farm in New Jersey. When she was fifteen, the family moved to California, the place to be in the 1960s. At university she studied medicine but didn't have the heart for it. In the late sixties she went travelling instead and hadn't stopped since.

To record her time in Australia, Aline started a journal, her name embossed in gold lettering on the red cover.

Aline's journal:

I've been spending a lot of time researching information at the Mitchell Library re: Aborigines and their relationship to dolphin/whales. Fascinating what I've come up with. One legend says: Dolphin: first wife (woman). The first woman came from dolphin. I find that wonderful. It seems that women played a very strong role with regards to their ancient beginnings. The Sun women who helped create the earth — responsible for introducing life onto this planet, and yet a lot of these stories Aborigine women are not allowed to hear. Extraordinary, isn't it. Really want to become more involved with this ...

Went to see Jonny Lewis, always a nice visit. If all goes well I'll be going next month with a group of people to Perth and Albany to stage a rally and try and get the whaling station closed down. Haven't been part of an activist group before but this one is in a peaceful space but very constructive, and besides they're working in an area that means a great deal to me: my friends of the sea. It's a matter of karma, I guess. Doing my share.

Jonny received a letter from Jean-Paul Fortom-Gouin with a cheque for ten thousand dollars, about one year's pay for a senior manager. The Phantom had come through.

Jonny pushed out press releases and planned an activist newspaper under the banner of the Whale and Dolphin Coalition, an official-sounding group. In reality it was Jonny and a core group that

Tom Barber.
Photograph by Aline Charney Barber.

included Tom Barber, Aline Charney, Richard Jones and Californian Pat Farrington. Jonny had no time to think about what he was doing. He felt that time was running out for the whales and he was so focused that sometimes he forgot to eat. The days and nights took on an intense, manic quality. He ran on nervous energy fuelled by marijuana. Negative thoughts were banished. Whaling must be dragged out of the closet and into the open.

The press releases were grand statements. Sometimes Jonny would add the names of other organisations, including Friends of the Earth and Project Jonah, or make up a few. If Save the Whale Foundation or Whale Embassy sounded right that day, he'd put that name on the press release. He and his friends were stacking it on, making it look like there were numerous activist groups massing to take on the whaling industry. Jonny upset some of the other established groups as they hadn't given him permission to use their names, but he didn't care. He went ahead anyway. He wanted to create havoc, a media storm.

Jean-Paul appeared in person again. The Frenchman didn't mind who was trying to stop whaling or what they called themselves — the more the merrier. He opened a Sydney bank account with fifty thousand dollars.

Furious energy flowed from and around Jean-Paul and Jonny, and people were coming and going constantly. Tactics were fine-tuned; media campaigns launched; equipment lists written and re-written; a magazine, to be called *Echoes*, planned. Dates and timelines. Stephen Jones, a friend of Jonny's, had access to video equipment to film the campaign. And who was going to drive the equipment to Albany?

There was laughter within the hard work. Another friend of Jonny's offered his musical services to lure the whales away from the harpoons. Jean-Paul and Jonny called him 'Toot the Flute'.

Poster designed by Peter Wright.
Photograph by Jonny Lewis.

'Toot, toot,' Jonny said, and he and Jean-Paul fell about laughing.

They concentrated, stayed focused. So focused, that Jonny and Jean-Paul almost finished eating some muffins before they realised they should have been put in the oven first.

Aline's journal:

Jean-Paul is an amazing man: small in physical stature, enormous in power energy, a wonderful, sensitive man. He is the director of the Dolphin Institute in Florida (Delphinid studies) for interspecies communication. He came out to Australia originally for the whaling commission conference in Canberra (Panamanian representative) ... north 'French' African by birth

... Anyway we've developed a wonderful relationship. In fact it's his fault I'm so involved with this entire project. HA! Last week I decided to be practical, work with the Whale and Dolphin Coalition until their departure for Perth, and then continue on with working (a practical route), then go on up north. Well within two hours of connecting with him he had me quitting work and committing myself to this project for the next two months. Even a possibility of going out on the two Zodiacs to confront the whaling boats. Also, I'll be helping Stephen Jones with the video and Pat Farrington with the New Whale Games, plus a bit of cooking. Freelance human being. HA! Looking forward to the experience.

We must try and stop the killing of the whales in Australia.

Jonny and Jean-Paul, in a haze of strong French cigarette smoke, sat at the kitchen table in the studio sketching a script for the demonstration and direct action. The campaign was taking shape. They had the resources, the organisation, the tactics and the media, but there was an element missing: they needed international attention to serve up Australia as an example to the rest of the world.

Jean-Paul put in a call to Vancouver, the headquarters of Greenpeace. He wanted to talk to Paul Watson, a Greenpeace co-founder, but he wasn't there. Bob Hunter, another Greenpeace co-founder and its first president, took the call. Jean-Paul had heard of Hunter, a journalist, and read his articles about the Greenpeace campaign to pit an eco-navy against the Soviet whaling fleet. These people had captured international attention in the North Pacific by using their bodies as shields to protect the whales. In 1975, the Russian whalers had fired over the heads of Bob Hunter and George Korotva as they manoeuvred their Zodiac, trying to protect a sperm whale. The whale was hit but the images of this incident swept around the world. Bob Hunter called the tactic a 'media mind bomb', an act to capture the world's attention.

The Greenpeace label was exactly what Jean-Paul needed. These wild, long-haired sea hippies would add credibility and attract media interest for the Australian campaign. He explained the strategy to Bob. If Australia could be turned into a pro-whale nation, it would be an enormous coup. Everything was set up in Australia to take on the Cheynes Beach Whaling Company. The equipment, organisation and funding were waiting for the Canadians.

'Can you come?' he asked Bob.

'I certainly can, but only if Bobbi, my wife, can come too.'

'The tickets are on their way.'

The Hunters' role would be as experts, Jean-Paul told Bob, and they would also be a focal point for the media.

Bob and Bobbi were excited. They said it was the dream of all Canadians to visit Australia. Bob, as a nineteen-year-old, had once headed for Australia but made the mistake of stopping at Las Vegas. His money quickly disappeared, but this time he would make it.

He had been the promoter, the media man, before Greenpeace became the Greenpeace Foundation in 1972. The group had its origins in the Don't Make a Wave committee, named after a phrase in one of Bob's newspaper articles, formed to protest US nuclear tests. Bob had a vision of Greenpeace becoming international, and a campaign in Australia fitted his view of a global organisation.

When he and Bobbi got together in 1974, they moved away from the anti-nuclear campaign and focused on stamping out whaling. Bob became president of Greenpeace; his membership number was 0000. Bobbi was the treasurer; her membership number was 0001. She was organised; he led with action, ideas and words.

The idea of using Zodiacs — rubber inflatable boats with outboard engines — came to Bob in 1973 during the anti-nuclear campaign. French navy sailors used inflatable boats to run down and board the ketch *Vega* off Moruroa in the South Pacific, where a nuclear test was to take place and where Canadian Greenpeace member David McTaggart was beaten unconscious. Bob saw that the speed, stability and manoeuvrability of the Zodiacs could be used to bring human shields between the harpoon and the whale.

But by 1977, he had too much coming his way and he was on his way to burn-out. Everyone wanted a piece of him and divisions were emerging within the ranks of Greenpeace. The phone call from Jean-Paul came at the right moment.

As they flew to Sydney, the Hunters felt good about what lay ahead. They had both been on two major whale expeditions in the North Pacific, and Bob had been on several successful seal protests. They were confident they could move things forward in Australia. They saw success as the amount of public awareness that could be generated. They were realistic enough not to expect to save a lot of individual whales from Australian harpoons, but many could be saved in the long term if Australians could be persuaded to close the industry.

They had no idea how many people would be involved in the anti-whaling campaign, the level of organisation they would find, or the physical difficulties they would encounter. But somehow it would all work out. It was down to karma.

Bob Hunter, Greenpeace's first president, in Sydney.
Photograph by Jonny Lewis.

The Canadians were exhausted after twenty-four hours' travel across the Pacific, including aircraft delays and mechanical breakdowns, and they slept for what seemed like a whole day. Jonny's studio was now looking like a headquarters, with people coming and going, sleeping on the floor and holding meetings. He took Bob to buy wetsuits and wet weather gear. A Zodiac dealer provided two sixteen-foot inflatable boats said to be stable in almost any seas. The words Cetacean Coalition were stencilled one side, Greenpeace on the other.

Bob was a major asset. Now Jonny's Whale and Dolphin Coalition would be taken seriously. All the reporters had heard of Bob Hunter's direct, non-violent action against the Russians, and Australia had become the number two international target. This was big news.

The sailor warriors of Greenpeace's eco-navy were gaining celebrity status. All this talk about whales holding a special place in the ecosystem made people uncomfortable. It was a challenge to the notion that man was the master of the world and could take what he liked. Initially Greenpeace was looked on by the establishment as a left-leaning organisation because of its anti-nuclear testing stand, but the anti-whaling campaign against the Soviet Union showed that Greenpeace was primarily pro-environment.

There was talk of a film being made in Hollywood, with Robert Redford playing Bob Hunter.

'I'd prefer Woody Allen,' Bob told Jonny.

Bob's reputation was not the only asset he brought with him. He had spirit and energy and never stopped. Those who spoke to him felt changed and recharged. He had what Australians call 'front'. He headed to the core of a problem without deviation or hesitation. To the Australians, he brought a template for direct action — use your own body to protect the whales and risk getting hit by a harpoon. It was crazy, dangerous and threatening. It was also magic.

The truck in Sydney, painted and ready to go.
Photograph by Aline Charney Barber.

The first media stunt was a demonstration of the abilities of the Zodiacs in front of the Sydney Opera House. They zipped around the smooth waters of the harbour while the press took photographs.

Aline Charney, with her experience of running a galley on yachts, wrote lists of what they would need when crossing Australia from Sydney to Albany. Everything was carefully planned.

Aline's journal:

So much to do before departure Saturday. Galley food (Tom Barber and I will take care of that on Friday). Everyone is pushing themselves so hard. The energy is so high. The people are fabulous. I am not an activist by nature but this seems to be evolving beyond anyone's control. When you have a group of people who care and respect each other and are involved with something good and positive — and you are given the blessing of the Aboriginal people — and it's allowing you to even connect more with the whales and dolphins and the timing seems to be right ... you must go for it!! It's beyond my control at this point ... this voyage across the desert to Perth in the truck with Tom and Stephen Jones is going to be fabulous, I think.

Jonny Lewis, Jean-Paul Fortom-Gouin and Pat Farrington flew to Perth. Jonny wasn't exactly sure how it would all come together, but once in Perth he held a press conference. On the way in Jean-Paul had slipped him a cheque for ten thousand dollars and whispered instructions.

Jonny got to his feet holding the cheque aloft. He had the money right here, he said, to pay the whaling station not to kill the last female whale of the hunting season. The whaling company wouldn't have anything to do with them, but it was fabulous publicity. Jonny found it exquisitely outrageous. The campaign had gained a life of its own, and it would all come together in Albany.

Two policemen approached Jonny and asked, 'Who's picking up your dole cheque in Sydney?'

Jonny ignored them. *Focus,* he told himself.

He got hold of the phone number of Sir Charles Court, the premier of Western Australia.

'We're coming and we're closing down the whaling station,' Jonny told him.

Jonny Lewis, Pat Farrington and Jean-Paul Fortom-Gouin launch the campaign against Australia's last whaling station.
Photograph by Greg Weight.

8 : Tarred and Feathered : August 1977

The first job of the day at the *Albany Advertiser* was to sweep the leaves and possum droppings from the top of the typewriters. The room, without a ceiling, extended upwards to the roof and there was nothing to keep the heat in or, when the wind blew, the leaves out. Possums lived next door in the peppermint trees by St John's Anglican Church in York Street, Albany.

The desks of the editorial department were islands piled with newspaper cuttings and old coffee cups. The typewriters, once the plastic covers had been removed, gleamed. Pure mechanical beasts. No electricity needed. You could smash out letters on these machines using brute force, each word punched into thick, porous newsprint. Anything was possible with such a tool.

Editor Bob Young, a Canadian, ex-Hong Kong, ex-Kentucky Fried Chicken manager in Perth, prowled. 'We don't pay you to phone people who aren't there,' he liked to say to those who couldn't find a key figure for a quote.

The phone rang and I answered. It was Pat Farrington of the Whale and Dolphin Coalition in Sydney. She was friendly; her voice was full of empathy and she wanted to be helpful. 'We're coming to stop the killing,' she said. 'We will force debate on the issue.'

'Sure,' I replied.

'How do you think people will react?' Her voice had an edge, something unsettled, something unsaid.

'This is a town with a long history of whaling,' I told her. 'There are some who will support the closure of the whaling station but the majority will support their neighbours. Local jobs mean local people.'

'I mean, how will they regard us?'

'If you mean will they tar and feather you, or string you from the nearest tree, I don't think so.'

'Yes, that's what I mean.'

'We're fairly civilised, I think.'

'Thank you so much. We'll see you there at the end of August.'

The Esplanade Hotel on Albany's Middleton Beach had never had such a large, single booking: ten rooms for a group of wildflower enthusiasts. The letter complimented Albany on its natural beauty. The group, the letter said, wanted to see it all and asked for outside rooms, those accessible from the street.

Diane Tonkin and her husband managed the Esplanade for Swan Hotels. The Esplanade was *the* place to stay in Albany. It was close to town, right on the beach and anyone who was anyone stayed there. The accommodation was separate from the bars. Live music on Friday and Saturday nights packed the rear of the hotel. Locals mixed with backpackers stopping for a break on their way around Australia. The bar at the Esplanade was ten deep at weekends.

The wildflower enthusiasts arrived in a wave of chattering excitement. Diane knew she'd been had as soon as they came through the front door. It was obvious they had little interest in native flora. They were all hippies, with flowing dresses, bell-bottoms and long hair. Diane immediately made the connection between the whaling protests she'd read about in the newspaper and the new arrivals. It was sneaky, but there was nothing she could do because the deposit cheque had been accepted. The hotel had to honour the booking.

John Saleeba couldn't remember a time when there wasn't some sort of reaction to whaling. Mostly it was background buzz, nothing serious. The anti-whaling lobby was active internationally but not locally, at least not in Western Australia, but the International Whaling Commission meeting in Canberra had been a platform for protests and the noise was getting louder.

John was the son of Wally Saleeba, one of the original shareholders in the Cheynes Beach Whaling Company. Over the years, the Saleeba family had picked up more shares until they held almost one-third of the whaling company. John had studied commerce and law and, aged thirty-four, had become executive director of the company, in addition to his full-time role at the law firm Robinson Cox. At the end of each day he would call in at the whaling company offices in West Perth.

'In the afternoon I'd pick up the new set of mail, pick up the typed-up stuff, sign all the letters and dictate the next lot,' he says. 'It was a routine there for a long time.'

The policy was always to give a written reply, no matter how bad the letter. At one stage John got a flood of letters from children asking him how he could be so evil.

'Someone would write, saying, "You only kill whales to make lipstick", which is the sort of nonsense that was put about. You'd write back and tell them whale products are used for a whole range of things and that ninety-eight per cent of the whale is used. I don't ever recall anyone writing back to say, "that's news to me" or even, "you're still a bastard".'

The whaling company created an educational package and started its own media campaign. One advertisement won an advertising industry award but it was like spitting into the wind — a wasted effort. The campaign by the anti-whaling lobby was swamping the media in the eastern states.

In Western Australia, whaling was a part of life and there was little media coverage. People would tell John, 'You seem to have the conservationists under control. It doesn't seem to be a big issue.'

He would reply, 'Well, I don't think so, but thank you for your support.'

He noticed that newspapers in the eastern states, mainly *The Sydney Morning Herald*, had begun campaigning against whaling. They had a little emblem of a whale on articles about whaling. It was seriously focused. At least the local daily newspaper, *The West Australian*, didn't campaign against whaling.

The company knew, if the press reports were to be believed, that the protesters planned to bring their case to the gates of the whaling station. They needed more security and so a wire fence went up. Tour buses were regular visitors and the station was on the list of activities for family holidays. Of all the questions asked at the local tourism office, the whaling station was at the top of the list.

The company feared that the station would be damaged. 'We knew from advice and from other similar campaigns,' John says, 'that there was a group of people within the conservationist movement that actively damaged things, sabotaged things. Ninety-nine per cent of the people were protesters with genuine feelings and views, but there were a few I would call bad eggs. That's the group we had to guard against because they could do serious damage.'

Jonny Lewis and the crew at the Whale and Dolphin Coalition celebrated when they heard about the new fence.

'We've got them worried,' Jonny said. He was pleased that the whaling company had been forced to hide behind a barrier, but this was also an indication that the campaign could get dirty. The coalition was causing a splash and their activities were making other conservation groups uneasy. The other groups preferred community-based action to bring the politicians onside, turn public opinion and change government policy. They were unsure of what was happening with the coalition, and worried that someone might get hurt.

Friends of the Earth in Western Australia didn't endorse the planned direct action protests, but nor did it condemn them. The state organisation had been coordinating the local pro-whale campaign for years but it didn't want to be involved in public events that might turn dangerous. If they were involved in the organising, they would be held accountable.

The Whale and Dolphin Coalition announced publicly what they were going to do and spoke to Friends of the Earth.

Peter Brotherton told them, 'Fine, go ahead and do it. It's a free world. It's your campaign. We know you're passionate and we know a lot of our members will join you.'

In Albany, the town was officially pro-whaling but the public face was showing cracks. The Albany Conservation Society, after a long and vocal debate, passed a motion calling for an end to whaling within seven years. The society decided that not enough was known about the Whale and Dolphin Coalition's plans to be able to support them, but individual members could become involved if they wished.

Now established at the Esplanade Hotel in Albany, Jonny kept getting questions from reporters about the planned action at sea, about taking the fight to the whale chasers. He replied that the equipment was coming and so was Bob Hunter, the Canadian expert and Greenpeace founder.

The coalition's frequent press releases announced World Whale Day as being something special that year; a major event of international importance was happening in Albany.

9 : Peace Caravan : August 1977

The truck was named Deimiri (Whale Spirit), in honour of the indigenous Australians who told the old stories about the dolphins and whales. It was spiritual food for the journey across Australia; the Aborigines were with the anti-whaling activists in spirit and approved of their mission.

Bobbi Hunter soon called the truck the Tinfoil Terror because, from her viewpoint driving the back-up station wagon behind the truck, it tended to fishtail at low speed. The ageing vehicle struggled along, with a trailer attached and overloaded with gear marked essential for the campaign against the whaling station.

Tom Barber could see they weren't going to make the four thousand kilometres from Sydney to Western Australia. The heavy equipment, including a large geodesic dome, would have to go. He unloaded the mini-convoy of truck and ageing station wagon at Bathurst, barely two hundred kilometres into the journey, and sent the surplus by train back to Sydney. He kept the most important stuff

— the two Zodiac inflatable boats and the outboard engines. Without these there would be no protest.

Tom had celebrated his thirty-first birthday earlier that month. He was comfortable driving outback Australia. He grew up in small country towns — Wentworth, Bourke, Albury, Goulburn — and guessed that Albany, in that respect, would be familiar territory. Born in Mildura, he spent his early childhood in the bush, moving from town to town as his father, a school inspector, followed his job. His father always kept a good workshop and Tom spent hours building things.

He had found a love for animals on Sydney's North Shore. His uncle, Frank Purcell, the mayor of Mosman, was a director at Taronga Zoo and Tom spent Christmas holidays among the animals after hours when no-one else was there. He finished his schooling in Sydney and went to the University of New South Wales, graduating in architecture. He was currently in peak physical condition, having recently built a house, called the Glass House, at Double Bay in Sydney's eastern suburbs, with Jonny Lewis sometimes acting as labourer. Excess body fat had boiled from their bodies as they worked with large sheets of glass in the humid air of Sydney.

With Tom on the journey was the likeable American, Aline Charney. She drove the Ford 250 flatbed truck, with Tom acting as her navigator. The truck, in a former life, had carted bricks to building sites in Sydney. Now struggling on six of its eight cylinders, it was finding the long distances challenging. In recent times it had become more used to the job of carting the inflatable whale Willie to media events around Sydney.

Jean-Paul Fortom-Gouin had funded the expedition. When money was needed Jean-Paul delivered, but funds were still tight. Most of it had been used to buy the truck and the equipment, and the trip across Australia was being done on a strict budget.

Aline couldn't believe that Jean-Paul had flown to Perth with Jonny while she was nursing an ailing truck across a continent. The thought rankled further with each hour she juddered along. Jean-Paul was debonair and sophisticated. He was charming. He was committed to saving the whales and matched his talk with action. They had become lovers and yet here she was on this lonely road and there he was flying high.

The Tinfoil Terror carrying equipment for the Albany anti-whaling campaign.
Photograph by Aline Charney Barber.

In the station wagon the lean, bearded Bob Hunter christened the procession the Peace Caravan. The Whale and Dolphin Coalition had promised singing, dancing, meditation and games, a real Age of Aquarius happening, and that's what they were expecting in Albany. The Canadians were dazed and jet-lagged. In a few short days they had gone from the sophistication of North America to the sparseness of the Australian outback. No extra charge for dust, flies and kangaroos. Neither of them had a clear idea of how far away Western Australia was. When the Peace Caravan was preparing to leave Sydney, the hopeful Canadians had asked whether they would see kangaroos. Tom and Jonny assured them they would see plenty. Forget about dozens or even hundreds, you'll see thousands.

The Nullarbor Plain doesn't have trees; the horizon stretches into infinity. It's like being the last pea on a plate the size of a country. People, on the rare occasion they saw any, talked slowly, as if they had all the time in the world. Passing motorists would smile and wave, happy to see anyone.

The Peace Caravan pulled into a truck stop. Bobbi was tired after driving through the long, dark telescope of night and jumping every time kangaroos bounced out of the black into the headlights of the car. The roos were in plague proportions after an unusually wet

Camping on the way to Albany.
Photograph by Aline Charney Barber.

period had turned the desert green. Hitting one at high speed could wreck a car and kill the roo — not a good image for the leaders of Greenpeace — and Bobbi had become skilled at slowing down without a jolt, careful not to wake her sleeping passengers. It was a long journey and the vehicle was falling apart, but the sparse landscape was different to anything she had experienced. It was disturbingly beautiful; Bobbi loved it.

As honorary treasurer of the expedition, she also had to make sure there was enough money for fuel and repairs. She parked the vehicle and went over to the driver's side of the truck to tell Aline the budget could stretch to a cooked breakfast in the restaurant. Aline smiled. Beside her in the passenger seat, Tom yawned. Before they left Sydney, Bobbi had planned for them to eat on the run, keeping stops to a minimum. At the back of the Edgecliff studio, Tom had built a cab for the flatbed with hinged sides that could be swung out for food preparation. Supplies were low and it proved impossible to carry enough fresh food and water for the entire trip. They were all tired and needed to stop.

After quick visits to the toilets, splashing water on parched skin and brushing tangled hair, the activists settled into the cafe. The anticipation factor was high. This was going to be the *best* breakfast. They waited and waited and waited. The restaurant staff, eyes averted, ignored the newcomers. Bobbi couldn't get their attention.

These people must have heard about the Peace Caravan from radio news reports. The media had been doing follow-ups, tracking the progress of the convoy across Australia, as Bob hit the telephone at roadside petrol stops to talk up the campaign. The accents may have given them away. The staff didn't want to serve them.

Bobbi erupted. She rose from her chair, walked behind the counter and went into the kitchen. 'We want bacon and eggs,' she said, 'and, if you want, I'll make them.'

The startled staff looked up at the woman with the foreign accent standing in the doorway. Bobbi stared at them, turned and headed back to the table.

A few minutes later the bacon and eggs arrived. They tasted fantastic. Bobbi felt she could face another day dozing in the car. It was someone else's turn to drive.

No amount of good food could help Tom. Mechanical trouble with the truck was getting to him. His stress level was high. At one point he found himself under the vehicle trying to replace a differential, something he'd never attempted. He snapped. In Mildura he shouted at Bob when he found the Canadian had slipped off to the pub for a quick beer. One of the eight people in the convoy insisted on playing Bob Dylan songs on his guitar. Tom couldn't stand it. He grabbed the guitar and threw it as far as he could over some bushes on the side of the road.

The flatbed truck, its white cab adorned with a painting of a whale, was a nightmare. They would drive for a while and the truck would stop. They would coast down the hills, and push the accelerator up the hill to extend the range. They would drive a bit more. The truck would stop again. This went on for hundreds of kilometres. Tom stripped the engine

Nothing could help Tom Barber. The truck broke down several times a day.
Photograph by Aline Charney Barber.

and put it together again. He sprayed ethanol into the engine. The blowback took his eyebrows off. Nothing worked.

Aline had a theory. Talk nicely to the truck. Deimiri had feelings. Be calm. Shouting won't help. It only upsets her. She spoke in soft, encouraging tones and, eventually, Deimiri fired up again and kept going for another hour or two.

Tom checked again, shouted at the truck and gave up.

Aline talked quietly and off they went again.

Tom shook his head. He couldn't believe it. He didn't know what pissed him off more, the fact that he couldn't find the fault in the engine or that Aline seemed to be able to sweet-talk the truck into going again.

The last breakdown was almost the end. The engine caught fire at the Nullarbor truck stop and they all watched as Tom got the situation under control with a fire extinguisher.

When Bobbi turned to talk to Bob, he was gone. She looked around and saw him walking in the distance. When she caught up with him, he was picking up sun-bleached bones from the roadside.

'I've had enough,' he said. He was going home. It took Bobbi twenty minutes to talk him around.

When they hit Western Australia the soil turned a rich red. The truck broke down again somewhere near Esperance in the east of Western Australia and a farm manager stopped to help. He took them down a long driveway to his house, where his wife and children laughed at the accents of the two Canadians and one American. The farmer found an old rag stuffed in the fuel tank. This is what had been, intermittently, stopping the flow of fuel to the engine. He also organised someone to drive, this time in a more powerful and reliable car, straight to Albany, now a day trip away. Bob had to be there for a scheduled press conference.

Aline's journal:

Our crossing of Australia took approximately 5 days, Deimiri breaking down continuously all the way. Thankfully her breakdowns were in close proximity to a gas station or a part was close in hand, and gratefully we had Allan Simmons and Tom who without their help and energy we wouldn't have made it. The last 72 hours saw us driving continuously around the clock — we had a date to keep. Upon arrival, no time to rest, connected with Jean-Paul, Johnny Lewis and Pat Farrington who were here already. Set up a base of operations at the Esplanade Hotel which is by the sea and convenient for our Zodiacs.

Sleep throughout this entire time and up until a couple of days ago was whenever you could grab it, 15 minutes here, an hour there.

When they drove into Albany Aline noted that the town used the whale to define itself. *A whale of a town. A whale of a radio station.* The whale was a symbol and yet they were killing the whales. They loved the image but were killing the real thing. The contradiction got to her.

The locals liked the connection to whales and loved their town. It had the best harbour in the world, and visitors were dared to disagree. Cold winds and seas rolled in from the Southern Ocean to King George Sound, which led to a deep and safe harbour, Princess Royal. The next landfall south was Antarctica.

Albany related to the whale.
Photograph by Aline Charney Barber.

The other big thing that locals liked people to know about Albany was that it was the first settled town in Western Australia, not that upstart Perth, a half-day's drive up the highway. The Swan River Colony, now the state capital Perth came two years later. But being first didn't bring the tall buildings and commerce of Perth. Albany was a service centre for a rich rural hinterland, a place from which to ship wool and wheat; to shop for clothes, machinery and cars; to stock up with groceries; to deal with state and federal government agencies; to consult with accountants and lawyers; or to holiday when the inland areas baked during the summer.

And whaling was a major contributor to the town's economy. Depending on the year's catch quota, set by the International Whaling Commission, the chaser ships had about 180 operational days. Each whale they caught averaged six tonnes of oil. Oil production was between four and five thousand tonnes, and although the world market for sperm whale oil was notoriously fickle, with prices swinging wildly, annual production sold for more than two million dollars. Add by-products, such as whale meal sold as animal feed, and the Cheynes Beach Whaling Company was pulling in around three million dollars a year. Profits could be $300,000 to $500,000, while the financial benefit to the community through wages and servicing the industry was considerable.

Jonny Lewis and Tom Barber getting the Zodiac ready.
Photograph by Aline Charney Barber.

So the town was keenly anticipating the arrival of the Peace Caravan, the Canadian Greenpeace founders and the Zodiacs. There were many concerns about what was likely to unfold, but not some small sense of excitement too at the novelty of it all.

The advance party — Jonny and Jean-Paul — had talked up the arrival of the equipment, especially the inflatable boats, to the media, and the two men were relieved to see the truck.

A boat launch was needed to give the protest some credibility. The Greenpeace veterans were the obvious choice but Bob and Bobbi, while experienced in piloting a Zodiac across the waves under the noses of whale ships, knew nothing about assembling the things.

The Phantom, Jean-Paul Fortom-Gouin, at Middleton Beach, Western Australia.
Photograph by Aline Charney Barber.

Tom jumped in. He'd put one together in Sydney for a promotional spin in front of the Opera House. He took the sixteen-foot Zodiacs to Middleton Beach in front of the Esplanade Hotel. He and Allan Simmons, a bicycle mechanic, assembled all the parts and added a thirty-five horsepower Johnson outboard with a fifteen horsepower auxiliary to each boat. Their activity attracted a group of locals.

'I wouldn't go out there with a rubber boat,' said one.

'Why not?' Tom said.

'There are sharks, great big white pointers. They're huge, ugly and hungry.'

Tom had no idea Albany was known for white pointer sharks.

Trailing blood and oil, the whale chasers were sometimes followed from the continental shelf all the way to the station at Frenchman Bay by white pointer sharks. A good catch meant a backlog of whales waiting to be processed. The sharks would take a leisurely feed while the whales were tied to a buoy. The flensers handled only one whale at a time on the flensing deck and it was a race to see who got more of the whales, the flensers or the sharks.

Gordon Cruickshank, master of the *Cheynes III*, says, 'Some of the sharks we saw were monsters. I would say twenty-five feet or thirty feet, and that's a big shark.'

Bob and Jean-Paul motored one Zodiac across King George Sound to the whaling station. As they coasted close to shore, they hit a submerged pipe used to transfer sperm whale oil from the whaling station to ships for export. By the time they limped back to Middleton Beach and the warmth of the bar at the Esplanade Hotel, reporters already knew about the accident.

It was Saturday 27 August 1977.

The battle had started.

10 : At the Gates : Sunday, 28 August 1977

The locals told them they were nuts, but the activists piled into the two Zodiacs at Middleton Beach in front of the Esplanade Hotel. Bob and Bobbi made the crossing of King George Sound to Frenchman Bay thinking of sharks. The jaws of a great white would saw straight through a rubber boat, they were told, and even a steel-hulled vessel was no guarantee of safety. These creatures were monsters and the whaling station was where they got a free feed. But the protest at the gates to the whaling station was the campaign's major land-based set piece. It had been scheduled and publicised long before the Canadians arrived in Albany, and the Zodiacs were key tools in the campaign.

They scanned the water as they zipped along — no sharks — and waved a placard at the whaling station: SHAME. The Zodiacs landed them safely below the whaling station car park.

Many demonstrators were already there, including Jim Cairns, federal member of parliament and former deputy prime minister in the

The Zodiac passes in front of the whaling station. Bobbi Hunter holding the sign.
Photograph Jonny Lewis Collection.

previous Labor government. Pat Farrington was at the centre of the crowd of banner-carrying anti-whaling activists, while floating at the back was the life-sized blow-up plastic Miss Cachalot, a descendent of Willie. Pat was relieved to see Bob and Bobbi step onto the beach. The whaling station was a place of grisly horrors for her, the flesh-stripped bones of the whales a monument to genocide. Images of how they died flooded her mind and she cried. But she looked across at Bob, who sent a thumbs-up, and she knew it was okay to cry on national television.

Just as the Canadians arrived, motorbikes roared into the car park. Bob smiled; the Hell's Angels had supported benefit concerts for Greenpeace in the States. But this was God's Garbage, the local bikie group, who had limbered up for the ride to the whaling station with a stop at the Premier Hotel.

The whaling station in Albany on 28 August 1977.
Photograph Jonny Lewis Collection.

The press group sat on an old sand dune overlooking the car park. I stayed in among the crowd of demonstrators, along with the photographers and television cameramen, trying to snatch a good quote from the shouting. Local police stood guard at the entrance to the whaling station. On the other side of the wire fence, the whaling station workers were amused to see God's Garbage. Many of their members worked in whaling or had close connections to the industry. Now we'll see some action, the workers thought.

God's Garbage dismounted and approached the line of protesters, and Bob realised their intent. There was shouting, pushing and shoving. Protesters milled around, trying to avoid confrontation. The police kept their distance.

Pat made a speech urging the local citizens to look for alterna-

Jim Cairns, former deputy prime minister (left), with Jean-Paul Fortom-Gouin at the whaling station.
Photograph Jonny Lewis Collection.

tives to whaling, other ways to make a livelihood. Her words were drowned out by shouts.

Jim Cairns, a veteran of peace marches, tried to speak but was shouted down as well. The megaphone came to Jonny. He looked at God's Garbage handing out pro-whaling pamphlets in front of him and sensed the mood could turn ugly quickly. His one hope had been Jim Cairns; surely there would be no violence in the presence of such a senior political figure. Jonny didn't know what to say. A wrong word could set off the crowd.

Oh, shit, Pat thought. She had felt the vibes a couple of days earlier when she was running New Games for children in an Albany park. God's Garbage wasn't a welcoming committee. The word was they were going to muscle the protesters out of town. Pat had spent her adult life bringing people together, working to create the positive — as a student leader, then through her New Games and lately in Australia teaching cooperative games to senior executives. She had dealt with the fluid world of politics and the stiffness of the corporate world — she had employees of a state premier and a group of straight bankers rolling around on the floor together. But angry whaling company employees in the form of bikies?

She stood her ground. With a smile on her face, she urged the protesters to hold hands, form circles and wail for the whales.

'We shall ...'

'What about our jobs?'

'... overcome.'

Pat Farrington speaking to the protest.
Photograph Jonny Lewis Collection.

Cheynes Beach Whaling Company workers confront the protesters.
Photograph Jonny Lewis Collection.

Pat ignored the shouts. Everyone followed her lead.

'We shall …'

A whale's like a cow.

'… overcome.'

Kill more whales.

'We shall overcome.'

Several of the God's Garbage found themselves surrounded by singing, smiling faces — inside a circle of chanting pro-whale demonstrators.

'Sharks!'

The shout drew the crowd to the beach. But on closer inspection, the shapes turned out to be dolphins. The protesters saw this as a sign that the sea, or at least its mammal inhabitants, was on their side. As they stood on the beach watching the dolphins, a rainbow appeared in the sky: another sign.

The appearance of the dolphins and the rainbow massaged away the anger from the demonstration. The protest at the gates to the whaling station ended calmly.

The last thing Bob Hunter wanted was trouble. 'We are peace-crazed, non-violent types,' he told reporters. He said that the campaign against the Cheynes Beach Whaling Company depended a lot on the weather. A gale would stop them but large waves wouldn't. In the north Pacific, Greenpeace had been able to keep up with the Soviet fleet despite a three-metre swell. The Zodiacs zipped along at fifteen knots and were more manoeuvrable than the chasers. Off Albany, he would be satisfied if Greenpeace could save the life of one whale.

Jean-Paul used his money to buy scrimshaw — carved whales' teeth — from the little shop at the whaling station. 'I'll give you a show,' he told the reporters.

He walked across the beach to the water's edge and threw the scrimshaw into the sea.

That same day a ten-year-old local boy went missing from Middleton Beach. He had gone to meet a friend in Marine Terrace, one street from the beach, and disappeared sometime that Sunday afternoon. The bars of the Esplanade Hotel emptied as locals and anti-whaling activists worked together to search for the boy.

Bobbi Hunter had seen a couple of boys down near the water, yelling at the protesters from rocks in the corner of Middleton Beach. Local children had been told these out-of-towners were the enemy, but she told her fellow protesters that a missing child touched them all and it was important to help at this time of crisis.

'We're going to stay and keep searching for that boy as long as the locals keep at it,' she said.

Bobbi felt that community attitudes, at least from those not directly involved in whaling, softened towards the protesters when they saw their efforts made in the search.

Jonny picked up a message from the front desk at the Esplanade Hotel.

'Mr Lewis or friends. Police were here this morning to see if you have had the film developed of Middleton Beach area, to see if anything shows up of the missing Albany boy. If you would, please contact police as soon as possible.' But the photographs revealed little and the search was unsuccessful. It was presumed that the boy slipped on wet rocks, drowned and that the currents took him out to sea.

As a child, Keith Forde looked forward to fishing at the whaling station with his dad. His father would monitor the two-way radio broadcasts and work out when the chasers would return with their catch. Father and son would drive from their farm and arrive as the chasers dropped off their catches. The blood and oil attracted schools of herring, and they would sit on the jetty in the evenings filling buckets with flapping fish.

Now Keith was twenty-nine and had lived in town for eight years. Many of his friends worked at the whaling station. His father and Ches Stubbs were mates. Once Keith went out on a chaser to see for himself but was shocked by the blood and guts, the cruelty of the killing. One old bull whale took four shots to kill. That left a bad taste in his mouth. Later, flying with his friend Ray Robertson, a spotter pilot, Keith saw the vast pods of sperm whales off the coast. From the air he could see the diversionary tactics taken by the whales and was impressed at how smart they were.

Keith took a stand, openly opposing whaling. In the small community, he was branded a radical greenie and was personally attacked for his views. He attended a meeting at the Albany Town Hall where Jim Cairns spoke against whaling, though his politics were opposite to those of Cairns. Keith felt keenly for his friends who worked at the whaling station and he argued a phased closure to protect those with jobs and families. It was the first protest he'd been involved in and his farming background made him stand out. The Forde farm was near that of Peter Drummond, the local federal MP, who spoke in favour of the whaling industry in Canberra. The families were friends.

Keith went down to Middleton Beach as the protesters prepared to launch the Zodiacs to take the battle to sea. He supported the cause but thought these people were crazy — they either had guts and conviction, or they were stupid.

Sir Charles Court, the Western Australian state premier, viewed the activities of Greenpeace with contempt. 'I don't think they deserve any public sympathy at all,' he said. 'The Cheynes Beach whaling project is one that has been conducted strictly within the rules ... the company is harvesting whales, not pillaging them as these people would have us believe. As far as I'm concerned, the Cheynes Beach people have always conducted their operations properly. It ill becomes a group like the Greenpeace people to come here trying to make trouble and to attack a company which has done the decent thing all along the line.'

11 : Bloody Compass : 1 September 1977

Jonny gave the compass a shake. The needle wouldn't do what it was supposed to. It was spinning wildly, pointing everywhere and nowhere in particular. He realised that he only knew about compasses as a concept. This was the first time he had tried to read one.

He didn't know much about boats either. He was more a surfer, getting his excitement riding the breaks up and down the east coast of Australia. He and Bob had bought the compasses from an army disposal store in George Street, Sydney. Bloody things, he hadn't realised the compasses might not work at sea, and now he found himself in the middle of the Southern Ocean in a tiny inflatable boat, heading in the direction of Antarctica with land rapidly disappearing behind him.

Opposite Jonny in the Zodiac was Jean-Paul. Both were wearing wetsuits under heavy-duty wet-weather gear. In the Zodiacs, you needed protection from the cold and rain, especially in the often wintry weather off Albany. They had both left their beds at two in the morning and launched the Zodiac in the dark at Middleton Beach to make sure they could slip in alongside a whale chaser as it headed for the sperm whale migratory path off the continental shelf. Around 5.54am the Zodiac picked out the *Cheynes II*, one of three whale hunters, and stuck close.

Jonny knew Jean-Paul would ask him.

'Give me a reading on the compass,' Jean-Paul said. His voice, as usual, was focused and clipped.

Jonny couldn't bring himself to tell Jean-Paul that he had bought the wrong type of compass. They didn't know where they were. We're stuffed, Jonny thought. We'll probably never see land again.

He studied the compass and noticed the needle slow a little in its wild spinning, so he quickly took a rough reading.

Jean-Paul, who had spent his life around boats, appeared satisfied with the numbers Jonny gave him.

Jonny caught a glimpse of the fifteen-metre-steel tuna-fishing boat that was shadowing the Zodiac as a back-up in case anything went wrong. It was the last time he saw it all day. The second Zodiac was out of action due to mechanical trouble. They were alone.

The sun came out, giving Jonny a clearer look at the chaser. He thought of them as whale killers. The crew were wild-looking men, capable of anything. The ship had a decayed look, peeling paint and rust. All it needed was a skull and crossbones flag to complete the menacing image.

'If we sink, I'm not getting on that ship,' Jonny told Jean-Paul. 'I'm going to swim and I don't care how long it takes.'

Jean-Paul was more worried about the engine breaking down and the boat drifting in the open seas. He knew the trip was going to be risky because the area was close to the big winds and waves of the Roaring Forties, but the weather was holding and the seas were kind. There were big long swells but it wasn't as rough as it could have been.

The *Cheynes II* as seen from a Zodiac.
Photograph by Jonny Lewis.

They had put a rough plan together in case of breakdown. If they couldn't get either the main or the auxiliary outboard to start, currents and weather willing they would drift along the coast and hit land in around two days. The support crew had instructions to search east if the Zodiac didn't return. The plan was more wishful thinking than a real chance at survival.

The other worry was that the flexible fuel tanks leaked, making the inflatable a floating bomb. Jean-Paul shouted when Jonny looked like lighting up a cigarette: '*No!* You can't light up in this boat.'

The tuna boat hired as back-up to Jonny and Jean-Paul was twenty nautical miles away and had no idea where the Zodiac was. The skipper of the tuna boat radioed the whaling station, concerned for the safety of the two men on the Zodiac. He didn't get a reply.

The Zodiac followed the whale chaser for hours, through endless sets of widely spaced waves, motoring up and down hills and valleys of dark-green liquid crystal. The cold, the wet and the motion of the small boat had an effect on Jonny's bladder. He was bursting. He stood up and peeled away his wetsuit. Nothing happened. He closed his eyes, breathed deeply and tried again. Nothing.

The steel-hulled whaling ship did a few tight circles then stopped in the middle of the ocean and the crew chucked a few fishing lines over the side. The captain leaned over the rail and joked, 'I'm terribly sorry, but we're lost.'

Jean-Paul, without hesitation, said: 'That's all right, we'll take you home.'

Kase Van Der Gaag, master and gunner of the *Cheynes II*, wanted the Zodiac, or 'rubber ducky', as his crew had named it, to follow him. There was something ridiculous about a rubberised open boat taking on his steel-hulled, ocean-going vessel. But Kase didn't underestimate the Zodiacs: they were seaworthy and they were fast. They could cause some real havoc if they got in among the three chasers when a pod of sperm whales was sighted. There was enough to think about without a bunch of protesters getting in front of a harpoon.

Kase had been at sea since he was young and loved the life, companionship and freedom. He was born in Rotterdam, Holland, in 1930 and went to sea in 1948. In 1969, he started at the Cheynes Beach Whaling Company and by 1973 he was a ship's master. At forty-seven, Kase was at the peak of his powers. His catch rates, the tally of sperm whales caught by the *Cheynes II* compared to the number of days at sea, were always consistent and good. He did his job well and got paid well for it. The killing wasn't pleasant but he loved being at sea. He had a good crew and a happy ship. They all worked together and looked after each other. Before joining the whale hunters, some of his men could have been labelled misfits on land but they'd turned out to be top seamen, people he was proud to work with. That was something that made the job special.

Kase knew he'd picked up the Zodiac as soon as he hit King George Sound. He had reporters on board and he guessed these would draw the activists to his ship — there was no point performing stunts if the media weren't there to record it. The newspapers were keeping a tally of the contest between the whalers and the protesters and Kase wanted a headline: WHALERS 1, PROTESTORS 0.

He didn't agree with the whaling company strategy on journalists. The company was keen to put its side of the debate and to do this they invited the media on board. He couldn't believe how stupid that was, it played right into the hands of the protesters. You don't take people to the meatworks to show them how their steak is made.

Paul Murray, a reporter with *The West Australian*, had managed to get himself and photographer Neil Eliot a berth on the *Cheynes II*. The previous night Paul told the protesters that he would be on board in the morning, knowing the protesters would then follow because his articles would be picked up by Australian Associated Press, the national news agency, and would flow to Reuters, the international news agency, guaranteeing global coverage. So it was no surprise when the Zodiac emerged from the lee of an island in the morning gloom. Paul found it exciting and he could see that Kase, the skipper, was enjoying himself, taking delight in leading the protesters all over the Southern Ocean.

Paul liked Kase. The skipper was pretty guarded in what he said but he was friendly. The deal with the Cheynes Beach Whaling Company was that the skipper and crew were not spokespeople for the company. They were not for quoting.

Some of Paul's media colleagues were not enjoying themselves. The cook on the whale chaser produced piles of greasy bacon and eggs topped with baked beans. One look and it was all over. Seasickness set in and stayed the rest of the day.

Paul was conscious of how quickly the situation could turn messy. The guys in the Zodiac were a long way out in an open boat. It wasn't a bad day, a bit lumpy, but the weather could quickly change for the worse. It was scary.

When the rubber ducky had begun following, Kase radioed the whaling station at Frenchman Bay, saying he was taking the *Cheynes II* due south. He didn't deviate from that course until he was far from the other two whale chasers and there was no chance the rubber ducky could break away and disrupt their hunt.

But the reporters were having a boring day. No harpoons. No whales. No confrontation. From their questions, Kase got the impression that most of the reporters and film crew were anti-whaling.

The *Cheynes II* monitored radio traffic until the inflatable lost contact with its support boat, the *Lea*. Kase knew, without checking his charts, that the compass readings were wrong — about thirty degrees wrong.

He decided to stay out overnight. He had fuel to last days. The protesters had no way to get to land on their own and they couldn't

Jean-Paul Fortom-Gouin at sea level in pursuit of the *Cheynes II*.
Photograph by Jonny Lewis.

last more than one day. He radioed the whaling station and received permission to stay at sea.

Kase saw the protesters more as an annoyance, buzzing flies, rather than a threat, but they were taking him away from his job and if it went on too long it would have an impact on his catch rates. The crew saw the Zodiacs as a direct attack on their livelihoods and those of the families they supported. These hairy bludgers were trying to take their jobs.

Kase had recently bought a shotgun, a pump-action riot gun. It was unlicensed, but it was such a good buy he couldn't resist it. He'd taken it down to the police station on Stirling Terrace where the desk sergeant had helped him fill out the gun licence application.

'And what's the gun going to be used for?'

'It's handy when the rubber duckies come around.'

The policeman stared at Kase and scribbled 'sporting purposes' on the application.

But as the day wore on, Kase decided against staying out overnight. There were no problems for the ship or crew, but he was worried about the reporters. They probably wouldn't appreciate a night at sea; they all wanted to return to port so they could file their stories.

He leaned over the side and shouted at the pair in the rubber ducky. 'Next time I'll leave you to find your own way home. I don't want you shadowing my ship.'

Jonny found it difficult to walk. It was dark, cold and miserable, and his muscles were cramping. There was no-one at Middleton Beach to help pack the gear. The whole scene felt surreal, like being inside a Salvador Dali painting.

He and Jean-Paul dragged the Zodiac up the sand, a strengthening wind hurling rain in their faces, and left the equipment where it was. On the way back to port, they'd stuck to the *Cheynes II* like glue. Jonny wanted to believe the chaser was heading home but he hadn't been sure until he got that first glimpse of land. They broke away as soon as they got that liberating fix on the coast. They hit the beach almost where they had launched in front of the hotel.

Bob met them as they approached the pub.

'We thought you were dead,' Bob told them.

The other protestors knew the back-up boat had lost contact with the Zodiac and had had no word on Jonny and Jean-Paul until they appeared on the beach in the dark. While Bob and the gang feared for their safety, Jonny hadn't known enough about the conditions off Albany to be scared. He had been locked into the protest and the day's journey, too involved in what was happening to speculate on the consequences.

Still wearing his wetsuit and wet-weather gear, he ordered two beers at the bar and headed for the toilet, his first civilised comfort stop all day. On his way, he saw Kase Van Der Gaag sitting at the bar. He nodded.

Kase smiled.

The toilet mirror showed a white salt film around a bright red face. When Jonny used the toilet, everything worked as it should have. He couldn't understand why he hadn't been able to empty his bladder at sea. He went to the bar, downed his beer and headed for a hot shower and sleep.

Kase had docked the *Cheynes II* at the Town Jetty about 8pm, jumped in his car, then raced around the hill to Middleton Beach and taken a stool at the Esplanade Hotel front bar. He ordered a beer and waited. The reporters who had spent the day with him had promised to buy him a drink. A whaling ship's master is a person of note in the town and word spread. The bar buzzed. The Esplanade was the unofficial headquarters of the protesters, so what was a whaling ship captain doing here?

The Esplanade had a good crowd. Albany was doing its bit to keep the national beer consumption curve heading skywards; it was possible to drink in a packed pub five nights a week.

When the two protesters, their sunburned, salt-encrusted faces glowing against their yellow raincoats, had banged through the door into the saloon bar Kase saw they were cold and tired. And he was sure they'd been shit-scared out on the open water without land in sight. He also knew they saw him and his crew as murdering monsters, rough and tough bastards, and they weren't happy about being led around by the nose all over the Southern Ocean.

The Frenchman with the beard, Jean-Paul Fortom-Gouin, came over and talked about the magnificence of the sperm whale. Bob Hunter also joined him. These two were natural talkers — using words as weapons. The Frenchman and the Canadian were in a small pub in a small town at the bottom of the earth trying to convince a whaling ship captain, born in Holland, that he was a murdering bastard and that what he did for a living was wrong.

The men were caught in conversation. Their surroundings faded as the words flowed back and forth. Not that there was much to see — the hotel was across the road from the beach but the bar faced the looming side of Mt Clarence. No water views. The local joke was that the building was designed for a wheat belt town but the plans got mixed up. Somewhere in Western Australia there was a pub overlooking a dry inland when it should've been overlooking Middleton Beach.

Jean-Paul told Kase that the sperm whale had the biggest brain on earth, perhaps in the universe. Who could say? Man knew nothing about how whales talk to each other. Scientists had shown they do communicate but the how was a big mystery. No-one had been able to decipher what they were saying. And how did they catch those giant squid? The squid were fast and the sperm whales relatively slow, so how did they do it?

Kase nodded at the barmaid. Same again. She was back a couple of minutes later with another round and collected the money from a pile in front of Bob Hunter. His shout. They talked and argued. Kase enjoyed himself. Bob Hunter was a nice guy. He meant well. They made a pact: Kase would stop whaling if Hunter could end the seal kill in Canada.

Kase could see Hunter's point. Sperm whales are magnificent creatures and Hunter held genuine beliefs, but what pissed Kase off was the way these people had picked on Albany, an easy target. These protesters, some of them probably professionals who made a living out of protesting, knew they could get good press coverage out of Albany. Good coverage meant donations. Thousands and thousands of dolphins were killed by tuna fishermen using nets and nothing was said in America because it was big money, but they could attack a small whaling company in Australia and get the reporters running. It was stupid.

But Kase knew the score from many years of experience,

whaling wasn't pleasant, but it was a job. He said, 'Killing is a not a nice thing and whaling is not a clean kill. You start chasing and the first whale is dead when you shoot it but the longer you chase them, the harder they are to kill and the more killer harpoons you have to use. In the end they get tough and the harpoons bounce off them. I had to use eleven shots once, on one whale. He wouldn't die.'

Paul Murray came into the hotel about an hour after Jonny and Jean-Paul. He wrote his article for the next day's paper on the way to port but he still had to find a telephone and dictate it to a copytaker on the other end in Perth. Neil Eliot, the photographer, had to make prints of the photos, take them to the Post Office and get a technician to make a connection to send the images via wire.

Paul saw Kase, Jean-Paul and Bob deep in conversation at the bar. He bought a beer and sat down at another table. The protesters were good drinkers. The deckies — the seamen from the whale chasers — were the same. Paul recognised a few of them scattered around the bar. They would stay at it until closing time. There was tension between the two groups and Paul kept an eye on what was happening, not sure how aggro the night would get.

Earlier that day Paul had leaned over the side of the whale ship to talk to Jean-Paul in the Zodiac. The Frenchman was so aggressive, so bloody bolshie, that Paul thought talking to him wasn't worth the effort.

But Bob Hunter was a different fish. Paul's whole story would centre on the Canadian. Bob was flamboyant, a real dope-smoking wild man full of derring-do, but he had passion for the cause and he was driven. Bob was the adventurer. Eventually he joined Paul's table. The Canadian wouldn't say much about the conversation with Kase, except that he thought Kase was a reasonable person and he had a lot of respect for him.

'Bob met the captain,' Bobbi says, 'they had a beer together. In the end, the captain said he didn't like killing whales but it was a job and jobs were hard to come by. If you were a whaler and if you had a heart, you didn't like what you were doing. It was a memorable event for Bob. You don't get the opportunity to talk to someone who's whaling.'

A storm hit town as Kase was heading home. He was glad he had returned to port. The guys in the rubber ducky would never have survived. Imagine what would have happened, what an impression that would have made on the public. It was already front page news. Losing two protesters would have been something else.

At the whaling station, nine whales were lined up on the flensing deck. It had been a good day for the whalers and, without a catch to drop off, Kase had been able to get to the pub at a reasonable hour.

Jonny was tired but satisfied as he rolled into his sleeping bag on the floor of the hotel room. What a day. Terrifying, exhausting, exciting. It had been the first day in Australian history that someone had harassed a whaling ship and kept the crew from their work; one-third of the whaling fleet was put out of action. Finally the campaign was hitting where it should — the whaling company.

12 : The Days Merge : September 1977

Bob Hunter became the front man as soon as he arrived in Albany. He debated with the whaling station spokesman John Saleeba on the radio; he spoke at the town hall; he rallied the activists from the saloon bar of the Esplanade Hotel; he took the battle to the gates of the whaling station and out to sea, and he did it all in the name of Greenpeace. It was important to him that Greenpeace open another front in the battle to save the planet. This would help realise his dream of a global organisation.

Bob attracted people. They loved his ideas, energy and movement, and he led by example. He would never ask anyone to do something he was not willing to do himself.

Aline's journal:

Telephones going, media reporters, TV, radio interviews, amazing press coverage, front page. This is a result of the experience Greenpeace brought to us via Bob and Bobbi Hunter, who came here from Vancouver to join us and allow us to benefit from their experience.

The campaign in Australia was going Bob's way, despite mechanical breakdowns, the cold, the weather and opposition from the Albany locals. He quickly established a relationship with the local media and favoured those who could get the message around the world, those with the international connections. He talked their language and was always one step ahead of them. Bob was the PR man for the whales. He unapologetically told reporters he was a traitor to his profession, that his journalism was built on opinion, not objectivity.

Each day Bob would turn something on for the media. He was their sort of guy, holding court in the bar, smoking and drinking, talking the quotable one-liner. He loved an unequal battle, tilting at windmills, taking the position of the underdog.

The West Australian, in a leader article on 31 August 1977, said there was no evidence to show that the sperm whales taken off Albany were dying out. They reported: '… it is the nature of things that the anti-whaling demonstrators will be able to claim a victory of sorts at Albany. They have done very little but gained a lot of publicity.'

Bob stuck to non-violent direct action. He had read the work of fellow Canadian Marshall McLuhan, who coined the term 'global village' to describe the fragmentation of humanity into an electronic tribal mind. From this, Bob created the media 'mind bomb'. This was what would work in Australia, he believed. Images and stories of dead and dying whales beamed into the homes of the people of the world via television, radio and newspapers would change opinion.

The Albany campaign was forcing the authorities into action: the police had to commit resources, politicians had to pay attention, and the whaling company had to meet the anti-whaling protesters on their terms. Bob kept them on the back foot. They couldn't escape the issue. Whaling was wrong — to the protestors it was like the buffalo in America, which had been so numerous when Europeans arrived that no-one could conceive of them ever disappearing.

But the big break came from the whaling company when they had allowed reporters on board to see for themselves. They thought the best way to handle the situation was to be open and show there was nothing to hide. Facts would swing the debate. The whaling company played by the rules and stuck to a strict catch quota system. The whales were killed quickly and cleanly. Almost every scrap of a whale's body was used, nothing was wasted.

Bob and Bobbi Hunter at the whaling station in August 1977.
Photograph courtesy *The West Australian*.

Bob and Bobbi laughed about the company's approach. 'All that happened was that you had poor whales being slaughtered and the reporters were horrified by it,' Bobbi says. 'That got out onto the wire services and everybody thought: You're doing *that*? Bob bought the public relations guy for the whaling company a couple of beers.'

Out on the water, the attitude of the reporters was not always consistent with what they wrote. Ship's master and gunner Kase Van Der Gaag could see expectation and tension in the faces of the professional observers and the sense of excitement as soon as the chaser got a line on a sperm whale. He guessed this wouldn't change the way they reported, but the same people who had told him they didn't like whaling certainly now had the scent of the hunt.

But Kase had to forget all that and switch his mind to the job of landing the whale. Nothing existed in the world but the gun and the whale. Shooting a harpoon cannon is not an aim-and-pull-the-trigger operation. The gunner must constantly assess and re-assess distances, allow for the ship's movement, judge the whale's pace. Now, Kase also had to make sure he didn't hit the rubber duckies or the protesters. The *Cheynes II* needed to be within a ship's length to get a clean shot.

'It's you and the whale,' Kase says. 'Nothing else counts. You can't say you aim like a rifle. It's like a shotgun. If you try to aim for a bird in the air, it's gone. It's not anticipation, and aiming doesn't help … I can't explain it.'

Kase liked to come on a pod of whales slowly and quietly. He felt that the whales listened. Full speed to half speed; slow to dead slow to stop. The *Cheynes II* could be steered for about two kilometres without using the engine. If you came in quietly, you could knock off a couple before they realised it.

Kase made the catch and moved the *Cheynes II* closer to deliver a killer shot. A rubber ducky was in the way.

He called to the protesters, 'The whale's going to die. You're making it worse for him. He's suffering. Move away.'

The inflatable motored away. Kase put in the killer harpoon.

John Saleeba, the spokesman for the whaling station, called Jim Cairns and asked for a meeting. They met in Dr Cairns' room at the Esplanade Hotel, alone, John sitting on one bed, Dr Cairns on the other.

The former deputy prime minister, who had also been a leader of the anti-Vietnam War marches in Victoria, wanted radical change in society. In 1976 he outlined some of his ideas in a paper called 'The Theory of the Alternative'. He wanted a cultural revolution to end oppression, exploitation, inequity and alienation. This resulted in the first Down to Earth ConFest in December 1976 in the Australian Capital Territory. More than ten thousand people — the subcultures of the hippie movement — came together in an alternative festival of yoga, politics, meditation, massage, alternative building, mud bricks, ideas, philosophy, music and vegetarian food.

At sixty-three, Dr Cairns was disillusioned with politics and was searching for a new way of life. He was crusading for an alternative Australia, and had become spokesman, the elder statesman, for the new culture.

He had bumped into Jonny Lewis at the Saturday markets in Sydney's Paddington a couple of months before and Jonny had asked him to help save the whales.

John Saleeba, however, saw Dr Cairns as a former deputy prime minister, a senior figure in society.

The executive director of the Cheynes Beach Whaling Company asked if it could be established that there are sufficient numbers of whales to allow for harvesting, would Cairns still have an objection?

John couldn't believe the reply he heard from Dr Cairns: 'We kill too many things. We shouldn't be killing whales.'

John knew the meeting was a waste of time.

On one point, John agreed with the anti-whalers. The whale was a wonderful creature, just as there were many other wonderful creatures, but he thought the whale was nothing more than that. He now understood that the anti-whaling people didn't want to conserve the species at sustainable levels so they could be a useable resource; they wanted whales protected, absolutely. He saw a parallel between the Hindu view of the cow as a sacred beast and the anti-whaling view of a whale. At least Hindus had a religious basis for their belief. John speculated what would happen if Hindu India achieved global economic supremacy and told the rest of the world what they could and could not do with cows.

He saw Dr Cairns once more after their meeting. The former deputy prime minister kept turning up at the whaling station demanding admittance, wanting to see for himself what was going on behind the fence. Eventually they let him in.

A group of families with children lived in cottages at the whaling station complex. 'He wanted to make a big fuss about how ugly it all was at the whaling station,' John says. 'When we invited him in, all the families came down, so he was surrounded by kids laughing and playing. So when they took a photo of him, he was in this lovely family situation with kids all hopping around. This blunted the impact, I think, of what he wanted to do.'

The Esplanade Hotel was booming. The bars were full of out-of-towners, travellers, the young, the middle-aged, politicians, experienced activists and starry-eyed newcomers. Lin Millington was the new receptionist, her first job since moving to Albany from Adelaide. She'd met a local man, Peter Pocock, and followed him west. Lin had done a secretarial course but had never been a receptionist before. Diane Tonkin, the manageress, told her to memorise the number of the local police station and call them if anything went wrong.

Lin was responsible for an old plug-in switchboard. When a hotel guest wanted to call Melbourne or Sydney they would lift the receiver in their room and give Lin the number they wanted. She would dial 011, tell the operator the number and ask for a ring-back price, which would then be charged to their room. The reporters lined up to phone their editors. Lin heard seven or eight versions of the same story each night as journalists dictated their copy before returning to the bar.

The protesters left each morning and returned late afternoon to take their places in a small bar near reception. There they rehashed the day's events and planned for the next day. They mixed with the press in the warmth of the parlour and saloon bars. The locals came in day after day at the same time, sticking to the public bar at the other end of the hotel.

The publicity surrounding the protests attracted more recruits and the booked rooms were now filled beyond the number of beds.

'The rooms were trashed,' Diane Tonkin says. 'I hate to think how many people were in each room but there were a lot more than the two they were supposed have. People were sleeping on the floor and there were clothes and rubbish everywhere.' The housemaids complained they couldn't clean the rooms properly. They didn't like touching a guest's personal things. But the hotel bills were always paid on time. 'They looked like they didn't have two cents to rub together but obviously somebody had the money for them. Someone was funding them.

'It was stressful trying to control everything. The reporters were there and the ones from out of town were trying to beat up a story on any little thing.'

When the bar closed at night, the remaining protesters slipped around the corner to a beach house, where they sat in front of an open wood fire, drinking and planning into the early hours of the morning. John Dawkins, a federal politician and later treasurer in the Hawke Labor government, had spent many school holidays at the weatherboard cottage. This time he was supporting the anti-whaling campaign. He was keenly aware of divisions in the town and the depth of feeling about the protesters.

The days merged. The cold, rain, wind and sheer hard work took their toll. Many of the early arrivals, filled with excitement at the prospect of fighting for the whales, drifted away, but the core group remained. After the booking at the Esplanade Hotel ran out, the protesters were told no rooms were available. All they could find was a shack in a caravan park at Emu Point. But there was no talk of giving up. They had an innocent, pure belief in what they were doing. The whales would be saved no matter what.

The men would get up at two or three in the morning each day to follow the whaling ships. Everyone would drag themselves from their warm sleeping bags to help get the boats in the water.

Aline's journal:

We've all worked so hard, driven ourselves almost to the point of exhaustion, physically and mentally, overcome amazing odds, lack of experience, machines not operating ...

Emotions have been operating on the highest levels: interpersonal relationships, interactions, like being with a group of people on a boat, living together 24 hours around the clock, exposed to the elements of Mother Nature.

Aline admired the courage Bobbi had shown in the north Pacific, where she had taken to a Zodiac to confront the Soviet whaling fleet, the first woman to use herself as a human shield for the whales. Aline consciously chose a support role for herself, however, rather than a front-line position in the Zodiacs. She realised she was in love with Australia. She was in the country on a tourist visa, and if she got caught up in anything unlawful it could affect her plans to stay. Her visa could be revoked at any time. She decided her best contribution was to ensure that those going to sea — Jonny, Tom, Jean-Paul, Bob and Allan — were looked after. She could keep the front-line troops healthy and see that everything ran smoothly. She was there when the Zodiacs went out and she was there when they came back.

Aline Charney in Albany, Western Australia.
Photograph by Tom Barber.

She thought of the protest as a homespun operation, spontaneous and driven more by the heart than the mind. Those involved cared about the whales. They wanted the killing stopped. It wasn't like the movies with heroes and heroines. Conditions were hard and planning was more seat-of-the-pants than textbook. There was nothing slick about it.

Aline's journal:

I'm realising more and more the state of some people: self-involvement, not being able to deal with situations or life in general outside of their protected, secure, patterned world. Others performing far and beyond the call of duty, selflessly and tirelessly working not just for themselves but for the overall good, for the whales, for our friends the porpoises, for the planet.

Aline sat at the kitchen table in the shack at the Rose Gardens Caravan Park, Emu Point. The others — Jean-Paul, Jonny, Tom, Stephen and Allan — were asleep. Pat had left for Perth to stage a one-person hunger strike outside the offices of the Cheynes Beach Whaling Company. Bob and Bobbi were at a motel down the road and hadn't told many people. They were worried about their safety. The crew had stopped wearing their Save the Whale buttons and T-shirts.

Aline understood that people were concerned for their jobs and incomes. She knew good people lived in Albany, but there was ignorance, fear and a lack of compassion for other forms of life.

Aline's journal:

WE SAVED THE LIFE OF 3 WHALES THE OTHER DAY...

In approximately an hour we'll be up and out the door for another round with the whaling boats. The weather is making it possible. This is what we've been working towards: taking the Zodiacs out, interfering (non-violently) with the chaser boats, getting in the way of the harpoons, making our presence felt. Winter is a very difficult time weather wise for this type of operation. The elements are against us. Overall, we're not as together as we should be to deal with these seas. And one thing I've learned: to deal with Mother Nature, to deal with the oceans and seas, to confront it, to relate with it, must be from a position of respect.

Jean-Paul and Jonny were able to go out 50 miles (crazies that they are, Jean-Paul specifically) and prevented the killing of three whales and stopped one chaser boat from killing at all by making their presence felt and seen. There has been a tuna boat taking media people out and giving us protection plus coverage. Protection is a strong word but in this case applicable (strange as it seems especially in the case of this group, probably the most non-violent, non-political activist group around, more of an affair of the heart).

Jonny today was refused gas at a station because a man saw his whale button on the front seat. Accommodation has almost been totally closed to us. Absurd, you say. Crazy. Yes, yes. You would think we were the SLA or some anarchist group of crazies instead of people just trying to save the lives of the greatest living species of mammals on the planet today.

The irony of all this is that Albany is a town which makes its rather prosperous income on tourism, using the whale as its symbol, as its logo, as its emblem. And yet they are the ones who are slaughtering them. Strange, isn't it. How much lovelier it would be if people could come here to see the whales to appreciate this not-too-common sight, to communicate with them. The money would flow, people's livelihoods wouldn't be in jeopardy and the whale could live. But the shareholders of Cheynes don't want that. All they can see is their profit of death. Money seems to do bizarre things to people. Or maybe I should say greed.

Bob Hunter called it shithouse karma.

He and Bobbi had to change their accommodation frequently.

'We were always on the run,' Bobbi says. 'God's Garbage, the local bikie group, wanted to beat the crap out of us. We didn't feel safe in any one locality.' The roar of motorcycle engines would crash into the country night air. 'It was intimidation. It was their job.'

One night the weather was so bad that Bobbi begged her husband not to go out in the Zodiacs the following day. 'You've done enough,' she said.

But Bob went anyway, but out on the ocean the mechanical curse struck again. Allan Simmons worked on the motor while Bob and Jonny sang. They didn't want to have to rely on the back-up motor in such rough seas. Somehow Allan got the main outboard going and they limped back to shore.

The mechanical problems, which had started with the road trip across Australia, continued. The outboards kept conking out. Tom and Allan kept pulling everything apart until they discovered the spark plugs weren't the right ones for the engines. They stopped going to that particular local outboard service shop and the problems stopped. Jean-Paul did runs to Perth to get parts.

Tom woke earlier than usual one morning, feeling uneasy. He went down to the beach. One of the Zodiacs was missing. Someone had either forgotten or been too tired to secure the boat properly the night before, or someone else had come along in the night and untied the boat. Tom jumped into the second Zodiac and started searching. He found the inflatable adrift in King George Sound about five miles away.

The work at sea was dangerous. Bob had wanted a back-up boat shadowing the Zodiacs at all times. Initially they had a steel fishing boat, the *Lea*, but that had dropped out of the campaign because it couldn't keep up with the whale chasers and Zodiacs. They found another boat but it was hit by the mechanical curse. That left the campaign without a mother ship like that used in the north Pacific.

The Zodiacs were alone. There would be no rescue, at least not from the Greenpeace side, if anything went wrong.

Jonny's karma was tattered, stuffed. He found Aline's flute-playing grating. He wanted to grab the instrument and smash it over his knee. One of the other women in the group, Rosie, was good at massages. That was more useful. A priest tracked Jonny down. He was checking on Jonny on behalf of his mother.

One of the reporters asked Jonny, 'What does your father think about what you're doing here?'

Jonny had no idea. He hadn't talked to his father, the former New South Wales premier, for months and his approval or otherwise wasn't an issue. Nothing was going to get in Jonny's way.

One morning as they prepared to go to sea again, Bob threw up on the grey wooden boards of the Emu Point jetty. He was too ill to go out in the Zodiacs. 'Jonny, would you go out in my place?' he said.

'Sure.' Jonny had been due to sit out that day's contest with the whalers.

'On one condition,' Bob added. 'You've got to call yourself Greenpeace.'

'I'm Greenpeace,' Jonny said.

The Hunters, Tom, Aline and a few others went into town for a change in pace from the intensity of the days protesting and the nights planning. They were running on adrenalin. They needed a night out.

Entertainment options in the town of Albany were limited and fast food in its infancy. The choices were Chinese takeaway, chicken dinners, pizza, or roast lamb at the Wildflower Cafe. For the locals, television gave some relief on the one or two nights a week when the pubs were quiet. The variety program, the *Don Lane Show*, had started. The saucy soapie *Number 96* kept people in their armchairs and gave them talking points for the next day. Sunday night the locals stayed home, exhausted, in front of the TV, watching the weekly movie.

The activists found a sleepy pub and settled for that ancient form of entertainment, conversation. Beer was the pub drink; a bottle of wine was out of the question. At Mt Barker, north of Albany, the wine industry was quietly building its market presence and honing its expertise. Plantagenet Wines had a cellar door fronting the Albany highway, but in the hotels there was little chance of ordering good wine over the bar.

Bob was animated and spoke in a mischievous tone. He talked about the Quakers, pacifists who believed in bearing witness, an act that could bring about change. This was part of what they themselves were doing in Albany — being witnesses to something that had, until now, taken place over the horizon, far out to sea and out of sight. Most people had a mind-image of whaling being all about wooden ships, long boats and hand-held harpoons. It was courageous and romantic, the *Moby Dick* story. They had no idea whales were being hunted using modern ships, sonar, cannons and explosive harpoons. Their job was to get this image around the world.

A man had come to Bob's house one day and given him a book, *Warriors of the Rainbow: Strange and Prophetic Dreams of the Indian People*. He put the book away for almost two years but took it with him when he sailed on a rusting ship to Amchitka in the Aleutian Islands in 1971 on the first campaign by the group that became Greenpeace. They were trying to stop a nuclear test by the United States. Bob wrote a column in *The Vancouver Sun* saying the test could send a destructive tsunami wave across the Pacific. The slogan for the campaign became: Don't Make a Wave.

As Bob and his friends headed to the test site, the ship dropped down a steep wave and the book slipped from a shelf. Bob caught it and decided he better read it before anything else happened. The Cree believed that magical champions will walk on the earth when the planet needs saving. When the rivers are poisoned, the seas blackened, and the animals sick, the Warriors of the Rainbow will come to teach the tribes how to make the earth beautiful again. This prophecy was the basis for calling the Greenpeace eco-soldiers Warriors of the Rainbow.

As they sip beer, Tom considers Aline, sitting next to him. He likes her. She hangs in there, is adventurous and fit, and she can keep going on little or no sleep.

And his mind is with the Canadians on energy. He talks about his passion for harnessing energy from the sun, clean and everlasting, a real alternative to nuclear power. His thinking has been influenced by Buckminster Fuller, the American designer and inventor who created the geodesic dome. Fuller dedicated a large part of his life to an experiment: what could an individual do to change the world and benefit humanity? The big question was whether humanity would survive in the long term. What the world needed was continual creative leaps to counter the degradation of the planet. Tom thought that energy from the wind — a gift from the sun's gravity — was the way to go.

He told the group that Albany people were familiar to him. They were similar to the characters he knew growing up in Bourke on the Darling River, Albury on the Murray, and Goulburn on the Wollondilly River, where he learned to be self-reliant, inventive and adaptive to whatever happened.

Aline likes Tom's dedication and his practical side. He never gives up, he never falls apart and he believes in what he's doing.

Aline had sailed a yacht across the Pacific to Australia via Tahiti, Fiji and New Caledonia. In a yacht there's a thin, artificial layer holding back the sea. She could sense the power on the other side and it made her feel small and humble. She speculated on what would happen if the whales ever got pissed off with humans, if they ever decided they'd had enough of humankind. The largest mammals on earth could place the ocean off-limits.

The group's entry to the pub hadn't gone unnoticed. A drunk and angry man broke a beer bottle and waved it under their noses.

Aline couldn't believe it. You've got to be kidding, she thought. We're no threat. We're nice people. The closeness of the threat brought her mind to the work of the whalers. They could flense a big whale. How long would it take to do the same to her?

Tom asked himself: What would Ghandi do? How would the Indian master of peaceful resistance deal with this? Bob had spoken of Ghandi and how the great man not only used non-violence but also knew how to get his message across. Bob and Jean-Paul have also been getting Tom to study tactics by reading an ancient Chinese text, *Sun Tzu on the Art of War*. Greenpeace was based on the principles of war, but Bob and his green warriors don't advocate violence. It was active resistance with a Ghandi-like philosophy; more peace-crazed than throw yourself at the barricades.

The protesters keep their eyes averted. They don't speak to or acknowledge the man holding the broken bottle. They make their way to the door, to the car and back to the shack.

13 : Harpoon : 5 September 1977

The two Zodiacs went straight to the continental shelf on their own rather than tagging along beside the whale chasers and found them by heading toward a tiny smudge of smoke on the horizon. The whalers could neither shake the Zodiacs nor lead them around the Southern Ocean chasing shadows. The chasers were among a pod of sperm whales.

The crew of the *Cheynes IV* bagged their first whale at 10.45 am. At the same time a radio message advised that the Greenpeace protesters were in the area. Under the command of relieving skipper Paddy Hart, the ship had left port at 4.10 am and headed south-west from Bald Head until they reached the edge of the continental shelf.

At 8.30 am Mick Walters, the operations manager and chief pilot, radioed that he had spotted a pod of 120 whales travelling south-west about forty kilometres south of the Warriup Hills. A sperm whale pod travelled between 3.5 and 4.5 knots per hour and

Mick easily calculated where the chasers should head to intercept the pod.

The three chasers hit the pod at 10.50 am, south of Waychinicup.

Alastair Anderson, in the crow's nest, spotted the rubber ducky twenty-five minutes later. There were two in the dinghy: the one with the beard was driving and the other, a big man, was sitting forward. They were having engine trouble. The inflatable stopped. The big bloke worked on the outboard and they got going again. Alastair shouted to the bridge below. The rubber ducky was about two kilometres away and approaching at fifteen knots.

In the air, Mick Walters watched what he described as a 'foldboat' — a Second World War collapsible dinghy used by commandoes — zigzag across the bows of the *Cheynes IV* as the ship pursued a bull sperm whale. He radioed the flight services unit at Perth Airport with a request that they advise the marine operations centre in Canberra that he had a complaint to make.

On the bridge, mate Gary O'Neill was concerned. Paddy had signalled from the gun deck to stop engines; the chaser could still move for a mile or more without power — the usual tactic when approaching a sperm whale. Gary saw the rubber ducky disappear behind the flare of the bow. He knew the dinghy had been having engine trouble and he didn't want a collision.

Paddy Hart had worked hard to become a skipper. His memories of his childhood in Ireland included the three years his father, a boiler attendant, had been without work. He was not yet fourteen when he left a Christian Brothers school in Dublin. He worked with his mother for a while, then trained as a chef in hotels, and at fifteen he went to sea as a galley boy cleaning pots and pans. When his ship sailed into Albany, he wanted to stay. He loved the place. It was a small country town and the Australians were terrific. He jumped ship, the *Irish Spruce*, in 1959, a few days before his twenty-first birthday.

It was easy to find work but harder to get paid. He got a job clearing bush land, but after ten to twelve weeks of backbreaking work, the contractors disappeared with his pay. Another contractor dobbed the Irishman in as an illegal immigrant and he was sent to jail at Fremantle, Perth's port, to await deportation on the next available British ship. The Irish consul intervened and got him a permit to stay, as long as he stayed out of trouble. That was eventually converted to permanent residency.

One night at Albany's White Star Hotel he heard two drunken deckhands from one of the chasers complaining about the state of the food on board. As a result he cooked for a couple of weeks before applying for a job as a deckhand.

He worked with a league of nations. The skippers: Reo Simojoki, a Fin; Axel Christensen, a Dane; Gordon Cruickshank, a Scot; Kase Van Der Gaag, a Dutchman; Ches Stubbs, an Australian. He made progress through the ranks to mate, first mate and, in 1975, he gained his skipper's ticket.

Kase Van Der Gaag helped Paddy when he was studying for his second grade skipper's ticket.

'He allowed me to take charge of the ship under his watchful eye,' Paddy says. 'Consequently, when the job came up, I actually had a bit of extra training.'

Sperm whale catching, in Paddy's view, was close to farming. It was a different story with the humpbacks. They'd been decimated. But with sperm whales, the females would generally be left alone because they weren't economically viable.

The first time Paddy saw a whale kill he was sad, but he found the hunt exciting. He couldn't believe such a large animal could be killed so quickly. It was in everyone's interest to go for a clean, quick kill. It was harder to put down a wounded animal that was trapped on the end of a line than one that was swimming free. It took longer and cost more and it wasn't a nice sight. Better to do the job properly in the first place.

'But unfortunately you were in a situation where things changed from minute to minute because of sea conditions,' he says. 'Sometimes, through no fault of your own, you'd only wound one that would require killer shots.'

Paddy's thirty-eighth birthday was three weeks away. He had the responsibility of a wife and five children and he was fighting for his job. It was a war of wits. He worked new strategies to keep the rubber duckies off-balance. Each time the protesters learnt a new trick Paddy would find a way to counter it. The decoy tactic used by Kase Van Der Gaag of the *Cheynes II* wouldn't work a second time because the protesters, now deploying two boats, would be unlikely to follow a single chaser. They would split up and each tag a ship.

By observation, the protesters knew that a chaser was about to change direction when they saw the smoke from the funnel change colour from white to black. This meant the engines were being pushed harder and could indicate the chaser had found a whale. Paddy saw the anticipation in the rubber ducky responses and deliberately steered away to put them off the scent.

He kept the *Cheynes IV* directly behind the whale as the ship tracked it by sonar. This time he decided not to come about when the whale came about, giving the protesters a false impression of where the whale was. As soon as the sonar operator indicated that the whale was coming up, and coming up fast, he kept the ship wide of where the animal would hit the surface.

At the last moment, Paddy turned the ship.

The whale surfaced.

Paddy fired the cannon.

The thud of the harpoon firing was an assault on Tom Barber's ears. The sound shocked and bruised. Time stretched like a tape recorder on slow speed. The world around him detached, was drawn out, elongated. He looked up and saw the harpoon fly towards the whale. From his viewpoint, sitting at sea level, the harpoon looked like it was coming right over his head. He couldn't understand that this was happening to him. This must be what an out-of-body experience is like, he thought. He didn't feel scared. He was in someone else's novel, a dream; he was somewhere beyond fear and death. There's nowhere to hide in a rubber boat. You may as well be sitting naked on top of a pole.

Tom saw the forerunner — the rope from the harpoon — smack the water a few metres from the Zodiac. There was nothing he could do to stop the momentum of the inflatable. It ran over the rope and tangled in the outboard propeller. He felt the Zodiac jerk sideways. The whale was dragging them down. This is it, Tom thought. We're going to die.

Jean-Paul shouted, 'Get the motor up! Get it *up*.'

Tom reached over the side and pulled the pivot pin on the outboard.

Paddy was relieved it was a clean kill. The whale had died instantly. If it had had any life in it at all, the line would have pulled taut and ejected the protesters from their boat.

'Don't you know what those signals mean?' he shouted, pointing to two black balls hanging from the mast.

'Yes, it means you've shot a whale,' came the reply from the rubber ducky. 'You've murdered a whale.'

Paddy had had enough. He got stuck into the activists. The two black balls meant the chaser was without steerage. The protesters were bloody idiots to get so close. Without power, the chaser had limited manoeuvrability; the ship was in a controlled drift and couldn't react fast enough.

Ali, the ship's cook, made a cutting motion across his throat when the rubber ducky passed. He'd made flour and chilli bombs. 'Burn their eyes,' he told the deckhands. Out on work release from prison, he was relatively new to the *Cheynes IV*. When he saw his skipper shouting at the protesters he grabbed a harpoon grenade and was preparing to lob it at the rubber ducky when Paddy rushed over and calmed him down.

Hurting the protesters isn't the best way to go, Paddy told him. 'Get back to your cooking.'

Alastair Anderson was always seasick. He was miserable most of the time. He figured he'd thrown up on every member of the crew by now, and he only stayed at the job on the chasers because he had a young family to support. He was usually the one in the barrel watching for whales because he had keen eyesight. He had the best view of what happened when the harpoon flew from the cannon.

Those blokes in the rubber ducky were lucky to be alive. None of the crew knew that Alastair had a camera with him. He snapped off a few frames before and after the harpoon flew. It had been pretty bloody close. If it hadn't been an instant kill, the protestors would be dead, and all they were worried about was the whale.

Jonny Lewis watched as the Zodiac containing Tom and Jean-Paul approached the shore at 3.45 pm. Jean-Paul, with his yellow knitted beanie and yellow wet-weather jacket, stood in the boat, waving to his fellow protesters.

'They fired over us,' he shouted. He was smiling.

A key element in the protest plan, discussed in Sydney before they arrived in Albany, was to get the issue before the Australian public. A court case was a perfect vehicle to debate the issues and having a local Australian involved would make an even better media story.

In Sydney, Tom had applied for a boat captain's licence, sat the exam at Circular Quay and passed, all in one day. Having a licence meant that the authorities couldn't claim Tom was inexperienced and try to stop him going to sea on the grounds that he could endanger himself or others.

Tom and Jean-Paul went to the Albany Police Station to make a formal complaint against the Cheynes Beach Whaling Company. Sergeant Terry Goodman did the interviews and prepared their statements.

Tom and Jean-Paul told the police that their lives had been endangered by the actions of the *Cheynes IV*.

'My purpose was to place myself between the whales and the harpoons to protect the life of a whale,' Tom said.

According to their statement, the harpoon from the *Cheynes IV* had been trained on the Zodiac as it zigzagged to put itself between the whale chaser and the hunted whale.

They recounted the events of that day. The two Zodiacs had found the three whale chasers about twenty-five nautical miles offshore, already in among a pod of sperm whales. Tom saw that one of the chasers had already shot a whale. He steered towards it but, again, he was too late to stop another whale being harpooned. Jean-Paul took over the tiller and Tom directed him towards the *Cheynes IV*, about seven hundred metres away.

As they approached the ship, Tom looked up. One of the crew waved an iron bar, then banged it on the side of the steel ship. 'Go home,' he yelled.

Another man pulled his penis from his overalls. 'Wankers. Wankers.'

Tom said, 'The whale that was being pursued by *Cheynes IV* was jumping out of the water. We then crossed behind and between the whale and the chaser. We then ran in front of the whale from left to right. The whale chaser followed us with his harpoon gun.' He estimated that the Zodiac was then fifty metres from the whale chaser and fifteen metres from the right rear side of the whale.

'We were about to cross behind the whale when we heard the thump of the harpoon discharge. The harpoon struck the whale near the centre of its back. The whale then put its tail in the air and dived. The rope from the harpoon struck the water about four feet from our boat. Our boat was under power and drifting towards the area between the whale and the whale chaser. Our boat then crossed the rope between the chaser and the whale, and our propeller became entangled in the rope. The rope lifted the motor on our boat and I had to release the pivot pin. The whale chaser was also bearing down on us.'

Jean-Paul said the harpoon 'shot across our bow and the harpoon cable slapped the water a couple of yards (two metres) ahead of us'.

That same afternoon, the Cheynes Beach Whaling Company wrote a letter of complaint about Geoffrey Thomas (Tom) Barber and Paul Gouin, also known as Jean-Paul Fortom-Gouin. It was addressed to the Officer in Charge, Albany Police Station.

Jean-Paul Fortom-Gouin and Tom Barber outside Albany Police Station after making complaints against the *Cheynes II*.

Photograph by Jonny Lewis.

Sergeant Terry Goodman interviewed Paddy and other members of the crew.

Paddy said, 'I would not have fired had there been any possibility of danger to those men.'

He described how he had backed off from firing the harpoon gun. 'As the whale surfaced, I saw the Zodiac coming from my port bow at right angles to the ship. The speed of the Zodiac would be about twelve knots. I saw that the course that he was on would bring him between me and the whale and across my bow. It was dangerous. I immediately uncocked the gun.'

The whale made a shallow dive and when it surfaced again Paddy saw the rubber ducky 100 to 150 metres off the starboard side.

'I cocked the gun and aimed it at the whale, which was surfacing off the port bow. The whale was about thirty metres on my port bow. It surfaced and I fired. The harpoon struck the whale behind the left flipper and it was killed instantly. The whale rolled over on the surface and did not dive.'

Paddy said, 'In no way could we endanger them by shooting above their heads or anything like that.' It was hard enough, he said, to catch whales under normal conditions. He didn't want to be the centre of an international incident. 'Protesting seemed to be in fashion. There was a lot of it about.'

The Cheynes Beach Whaling Company had told their men it was their duty to render assistance if people were in difficulty.

Two days later a CIB detective from Perth visited Jonny and Tom. The detective also wanted to see the Frenchman, but Jean-Paul was away.

The policeman said he was investigating the incident on 5 September in which Tom Barber and Jean-Paul Fortom-Gouin had said their lives had been put at risk by the whalers. He told Jonny and Tom there had also been an official complaint from the Cheynes Beach Whaling Company.

Jonny told the detective that they had a right to try to stop the killing of whales, and besides, the Western Australian police had no jurisdiction fifty kilometres out to sea, well outside territorial waters. In fact, the whalers had interfered with the legal rights of the protesters. Greenpeace had every right to watch the whales.

The detective was unimpressed. Their actions, he said —

zigzagging in an open boat in front of a whale ship — had endangered the crews of the whale chasers as well as themselves. He told them about a 1975 amendment to the Criminal Code which could be used to prosecute those committing a crime up to 160 kilometres offshore.

Jonny had heard of the amendment but didn't know the details. He had legal advice that state police had no jurisdiction. As the detective continued to talk, he began to feel they were in serious trouble. They would, if the policeman were to be believed, be arrested if they harassed the whale chasers again. All their equipment, including the Zodiacs, would be seized and impounded, to be used as court exhibits. Police would oppose bail and they could end up in jail waiting for their case to be heard in court.

Jonny got the message.

The detective told the protesters to end their activities in Albany. If they wanted to pursue their objective of closing the whaling station, they should do so by lawful and peaceful means.

Jonny and Tom were polite. They assured the detective they would think about what he had said.

Jonny worried most of the night. At ten o'clock on the morning of the next day he called the Albany Police Station and told the detective that Jean-Paul had left Albany in the early hours of the morning and that most of the protest group had checked out of their accommodation and were leaving Albany as well. Two men would remain behind with him to repair the truck and load equipment.

That same morning the police announced that the whaling company had been exonerated of blame. The complaint by the activists that the whalers had endangered their lives had been dismissed. The company had committed no offence and had taken all safety precautions. The police investigation found that a harpoon had not been discharged close to the protesters.

Jonny and three others were ready to leave Albany but couldn't

Whalers blameless. Tom Barber after police dismissed his complaints against the *Cheynes II*.
Photograph by Jonny Lewis.

find anyone to fix their ailing truck. One repairman said he could do the job but changed his mind when he found out the customers were anti-whalers. No-one would rent them a room and their booking at the caravan park was due to run out in two days.

Jonny had been rattled by the appearance of the CIB detective and he sought a lawyer's advice. The lawyer told him that under laws still in force, the Western Australian government's jurisdiction extended offshore only as far as a cannon ball could be fired. The Cheynes Beach Whaling Company operated fifty kilometres offshore, beyond the capability or control of government gunpowder. The 1975 amendment to the Criminal Code, referred to by the CIB detective, was a flawed piece of legislation. It was unlikely to withstand a challenge. It would, in all likelihood, be deemed unconstitutional, but the state government was due to change the law before the 1978 whaling season began to give police more power to stop the protesters.

Before leaving, Jean-Paul had said, 'I believe we have succeeded in making people think about whether they want whaling to continue in Australia.'

Jean-Paul Fortom Gouin at Middleton Beach.
Photograph by Aline Charney Barber.

'We changed the consciousness of Australia,' Bobbi Hunter says.

She and Bob went from Albany to Amsterdam, where the World Wildlife Fund had organised a phone-in fund-raiser for Greenpeace. They raised $150,000. The Greenpeace office in Canada was $150,000 in debt because of the global anti-whaling campaign. On the way home to Canada, Bob and Bobbi stopped in London, where Greenpeace had found a ship, the *Sir William Hardy*. Instead of paying off the debt, the Hunters used the money to buy the ship, which was later renamed the *Rainbow Warrior*.

14 : The Eye of the Whale : 22 September 1977

The rubber ducky appeared out of nowhere about 10am. No-one had expected to see the activists again; everyone thought they had left town. The Zodiac shadowed the *Cheynes IV*, skippered by Chris 'Axel' Christensen, until 12.45 pm, when the chaser sighted a whale. The inflatable zigzagged about seventy metres in front of the bow of the ship.

Tom was comfortable in the Zodiac; he had spent his teenage years in small boats on Sydney Harbour. It took three hours of motoring up and down the swell, the crests peaking every hundred metres or so, to get to the edge of the continental shelf. Tom kept the throttle of the outboard at an even speed. He didn't want to stress the engine. He wanted to get home.

Jonny was tense, keyed up. He still had difficulty relieving himself at sea. No such problem for Tom. He would peel down his wetsuit, hang off the back and be done in under a minute.

In the hours when nothing was happening, when they were on their way to find the whale hunters, Tom and Jonny chatted. A pod of what looked like a thousand dolphins appeared, stretching from horizon to horizon parallel to the boat. Jonny said it was a good sign. They drank warm Coke and ate chocolate. The bottom of the boat was an organic soup of floating bits of bread, chocolate bar wrappers and other scraps.

Tom Barber at sea in the Zodiac, chasing Australia's last whaling fleet.
Photograph by Jonny Lewis.

The Zodiac was battered. Whoever sat in front was sprayed with seawater. Tom preferred to steer, even though he'd lost his distance glasses early on when he was tipped into the water as the Zodiac changed course without warning. He wasn't going into the water again; better to keep Jonny away from the steering. Jonny was the front man, the passionate speaker, so involved in the campaign that Tom reckoned he must've been a dolphin in a previous life. In fact, if he squinted and looked at Jonny from a certain angle, he even *looked* like a dolphin, almost.

The sun is severe in Albany, even on cloudy days. Artists delight in the clarity of the air and the sharpness of definition, but those out at sea in an open boat are easy game for the harsh ultraviolet rays. Tom's face burnt and peeled no matter what he tried: a hat, sunscreen, pink zinc. In desperation he grabbed a chunk of greasy salami from the thick-cut sandwiches Aline had packed and rubbed it on his face. It didn't help. His skin still turned bright red and fell away in painful strips.

Once among the whales, Tom skipped the Zodiac across ridges of swell, zipping in front of the chasers, trying to spoil the aim of the harpoon gunner. He thought the whalers looked more organised, more determined this time and there was no abuse, no shouts of 'wankers', no banging the side of the ship. Tom thought, Shit, this is going to be a battle.

Jonny took photographs and watched the chasers, trying to second-guess their next move. One chaser was moving in on a whale. He motioned to Tom. The Zodiac took off.

They closed on the chaser. Jonny scanned the ship and saw the gunner tracking the Zodiac with the cannon. He's trying to scare us, Jonny thought. He turned away and looked down. A dark shadow in the water caught his eye. The water was boiling; a shape was rising.

'Tom! The whale's coming up beneath us.'

Tom throttled back and the Zodiac slowed. The head of the sperm whale broke the surface and a mushroom of water vapour spread in the air. The breath of the whale covered Jonny and Tom and their nostrils cringed from the stink. They put on more speed, trying to catch up, trying to keep the Zodiac between the whale and the whale hunter, but the cannon cracked the air and the harpoon hit home with a loud thud. The grenade exploded a few seconds later and blubber shot into the air.

Jonny was rattled. His hands shook. He had been sure the harpoon was going to hit the Zodiac. Tom was lucky — at least he could hold onto the tiller. Jonny recovered and swung his attention to the whale.

It had been hit. Jonny sagged. He felt impotent, useless, filled with overwhelming rage and frustration. Heavy duty. His focus was shattered, but a calm part of his mind, a splinter, noted the efficiency of the whalers. They put two more harpoons into the whale, making sure it was dead, brought it alongside, pumped it with air, put a flag on the body and disappeared over the horizon for the next kill.

Blood-red waves lapped around a creature as big as a house.

'What are we going to do, Tom?'

'Let's just touch him and say goodbye.'

Screeching seagulls swarmed over bloody bits of whale as Jonny rubbed his hand along the whale's side. He took the tiller and Tom did the same, also touching the eye of the whale.

Jonny was freaked out by the sudden violence, the immediate terror of the experience. His camera had come apart and he couldn't put it together because his hands were shaking so badly, but at that moment a photo didn't seem important anyway. For Jonny, the ride to shore took forever. Inside him, there was an emptiness surrounded by anger, and on top of that a black, deep sadness.

As the day grew dark, Jonny thought he saw a white dolphin, a ghost image.

Jonny slipped out to the whaling station before he left town, mingled with the tourists and took photographs. He turned one of the images into a postcard and mailed it to each shareholder of the Cheynes Beach Whaling Company; their families would then know what they had invested in.

After he returned to the eastern states he staged an exhibition in Canberra called 'Slaughter Gala: Australia's Shame at Cheynes' using slides and photographs 'direct from the whale war in Albany'.

Tom and Aline stayed in Albany for a couple more weeks. The tensions of the campaign eased and locals started dropping by to chat about the issues.

Aline told them, 'When they close down, you guys are out of it. No job. No nothing. And maybe you've got to think differently.'

Rather than telling the locals they were horrendous, they talked about alternatives to whaling, including whale-watching. These people had families and children. Tom and Aline felt better when they left town, having had that time to connect, and they hoped that

Aline Charney and Tom Barber in the Tinfoil Terror.
Photograph Aline Charney Barber Collection.

the message had got through: they were not the enemy.

Tom and Allan Simmons took the truck's engine apart and put it back together. A few pieces were left over but the engine worked on all eight cylinders now. Aline, Tom and Allan took the truck, loaded with the Zodiacs, on a leisurely drive back to the eastern states. They camped out, enjoyed the sites, and Tom and Aline got to know each other.

Tom gave Aline a present, a branch from a eucalyptus tree. 'Use it to swish the flies away,' he said.

The protest was still making news across the world. During the Albany campaign, Greenpeace veteran Rex Weyler was on the *James Bay, Greenpeace VII*, in the northern Pacific, confronting Soviet whalers. When he returned to Vancouver, he stumbled across one more strike against Australian whaling.

Greenpeace had received a tip-off that a shipment of Albany sperm whale oil was in Portland, Oregon, on board the Norwegian tanker *Stolt Llangdaff*. US law prohibited the importation of whale products, and the ship was told to leave American waters. It turned up in Vancouver, where dock workers refused to touch the cargo. The 1,200 tonnes of whale oil left with the ship and dropped out of sight.

On 27 October 1977, *The West Australian* reported that the Cheynes Beach Whaling Company had sold the oil to a European customer at some prior time and the cargo was no longer the company's responsibility.

Being a small operation in a small town at the bottom of Australia wouldn't work as a passive defence anymore. The Cheynes Beach Whaling Company was being targeted on a world stage. Sperm whale oil buyers wanted reliability of supply as well as quality and price. How long would they continue to do business with the Cheynes Beach Whaling Company?

15 : School Project : 1977

Phoebe, an eleven-year-old boarder at St Margaret's School at Berwick in Victoria, thought a lot about her project. Whales fascinated her — those giant, sea-going mammals that suckled their young and had brains similar in complexity to that of humans — but she'd discovered something terrible. The animals were being hunted and killed right here in Australia.

She came home for school holidays and talked with her father about her school project. We shouldn't be whaling, should we?

Her father, Malcolm Fraser, the prime minister of Australia, didn't think much of whaling either. He didn't like the trade and thought it was something the world could do without. How to stop whaling was another matter. His philosophy was to do the right thing, make decent decisions, and support for those decisions would follow. You either had an instinct about what was right or you didn't. On the question of whales, with the scientific evidence available, there was only one answer. His farming history also influenced his interest in environmental matters.

'I think a lot of people associated with the land are concerned about the environment,' Fraser says, 'because, if your own particular environment goes to pot, your farm goes to pot. Your livelihood goes.'

Some years earlier, at meetings to start the Australian Conservation Foundation, Fraser had persuaded the then federal treasurer Harold Holt to grant tax deductibility for donations to the foundation.

Fraser could speak frankly with his daughter in private; as prime minister he had other issues to consider. He was aware of the growing campaign against whaling, the media attention and even personal representations from members of his own Liberal Party. His office had been flooded with letters. No other issue had ever created so much paper. It was an issue that needed careful handling. A ban on whaling could have political repercussions in Western Australia, where the Liberal premier, Sir Charles Court, would fight any decision meaning a loss of jobs. Malcolm Fraser could also face opposition from within his own cabinet.

The Frasers generally kept their children away from the media. But a chance remark would change that.

You can tell a man from the east, but you don't tell him everything.

State rights was a lively issue nationally and, in Western Australia in particular, Canberra-bashing had always been a popular sport. The eastern states couldn't be trusted when it came to the economic health of Western Australia, especially on the issue of jobs. The feeling on the streets was that Western Australia's north-west mining boom was propping up the rest of the Australian economy and the state was getting the bad end of the economic deal. Seceding from the Commonwealth wouldn't be a bad idea.

The federal government had the power to end whaling. All it had to do was withdraw the Cheynes Beach Whaling Company's licence. The national government had responsibility when it came to the International Whaling Commission, a multilateral organisation, but there were significant political dangers. The ruling Liberal Party in Western Australian was pro-business and whaling was a business. Any move by a government, especially the federal government, to close a business would get an angry response from the state Liberal Party leadership and membership. The government of Sir Charles Court didn't want interference from the federal government, even if it was of the right political persuasion. Blocking encroachment by centralist Canberra was a vote-winner in Western Australia.

The Liberal Party was then a loose amalgamation of independent state divisions, and Liberal state premiers were powerful. The state branch of the party had influence over the selection of candidates for federal seats. Liberal senators or members of the House of Representatives from Western Australia, supporting the closure of the whaling station, ran the risk of losing preselection and finding themselves without endorsement at the next election. They needed split personalities: they had to support the Fraser Government *and* be seen to be pro-Western Australia.

Peter Drummond, the federal member for Forrest, the electorate covering the whaling station, was popular among his Liberal Party peers in Canberra. Backbench MPs wouldn't want to see him disadvantaged or suffer the electoral consequences of the closure of a local business as a result of federal action from a government of which Peter Drummond was a member.

In Sydney, Chris Puplick wanted whaling on the national agenda. He had stepped aside as federal president of the Young Liberals, a role he had held since 1974. He wanted his political party to be on the right side of the argument and he knew there were few MPs who had an awareness of whaling as an issue. He decided to spend time at the Project Jonah office.

Project Jonah was a genuine grassroots fundraising and lobbying organisation, and at this level environmental issues were not seen as radical, or left wing, or the preserve of the Labor Party. Working from the Project Jonah office, Puplick found dedicated and smart people with terrific social and business networks. But, he says, 'They were perhaps a little naive in terms of who they needed to apply pressure to and how to approach them.'

He spent some time working through a series of critical questions: who could be influenced and how could they be influenced; where were Project Jonah's resources, its people, located; what did they do, who did they know? Effective strategies could be built on the answers.

But, as far as Puplick was concerned, Project Jonah's important asset was that it had a view about whales and didn't have a view on anything else. 'It wasn't a conservation group or an environmental organisation. It had a single focus. You have to have the discipline not to be drawn into other issues.'

Puplick asked for Project Jonah's view on uranium exports and Richard Jones, who used his copywriting skills to write advertisements for Project Jonah, thought hard before replying. He knew this was an important question.

'It's got nothing to do with whales,' he said.

That was the test. The anti-whaling campaign had a far better chance of getting the ears and support of the lawmakers and bureaucrats if these authorities could be sure that an organisation such as Project Jonah would not sideswipe the government on an unrelated issue. And the establishment felt they could support the cause: anti-whaling had become middle class and respectable. Politicians could support the campaign without fear, knowing the machinery of the anti-whaling campaign wouldn't be turned against them.

The bureaucrats in Canberra, who had essentially administered a quota system for whales under the umbrella of harvesting a natural resource, saw public opinion start to turn. The mood was pro-whale.

This wasn't an issue of stock depletion; whales were special and needed to be protected.

The delegation from Project Jonah — Joy Lee, Henrietta Kaye, Tony Gregory, Jennifer Talbot and Ken Harrison — was prepared for a tough time. They'd rehearsed their arguments. Malcolm Fraser had to be persuaded that whaling was something the community wouldn't accept. Getting the meeting was a major step. They were aware that jobs at the Cheynes Beach Whaling Company in Albany were at issue and a national election was to be held the next month, in December.

Fraser was in Melbourne to support Neil Brown, a member of the House of Representatives, in his electorate. Brown was one of several people including lawyer Ian Renard, adviser to Malcolm Fraser, and Senator Alan Missen, who had helped Project Jonah win a meeting with the prime minister.

The delegation launched into their pitch. They were aware they might never get a better chance.

'We started to explain our concerns,' says Henrietta Kaye, 'and he told us, "Please, you don't have to tell me. My daughter has already made me aware of the issue." '

But Fraser could not give Project Jonah the outright ban on whaling they wanted. He was concerned about the situation but had to find a way to overcome opposition from Western Australia. Fraser said he would create an independent inquiry into whales and whaling. It would not help the pro-whale cause if there were an impression that the prime minister had already made up his mind on the issue.

'We were not expecting an inquiry,' Henrietta Kaye says. 'We were there to put our case and to present him with 100,000 signatures on a petition which we had collected.'

The delegation gave the prime minister a book titled *Mind in the Waters*, by Joan McIntyre, Project Jonah founder. He thanked them; he would give it to his daughter, she liked whales.

Outside, the press was waiting for the Project Jonah delegation.

How long were you in there?

We were in there twenty minutes.

How did it go?

It went well. His daughter likes whales.

The press loved that story. The daughter of the prime minister was on the side of the whales.

'Phoebe spoke to me about it,' Malcolm Fraser says, 'and I mentioned her name in the meeting with Project Jonah and somehow that got into the press.'

A newspaper ran a cartoon with a whale holding a placard saying, 'Vote 1 — Phoebe for Prime Minister'. Phoebe was proud of that. She stuck a copy on the door to her bedroom.

'I think the throwaway line was picked up to make it seem to be more than it was,' Phoebe says. 'I do feel that a lot more was made out of it than it should have been. If the truth be known … I'd like to say it was all me. However, for me it was all about a school project. The credit goes to Dad.'

During 1977 and 1978, the prime minister's office received more than 37,000 representations calling for an end to whaling. Seventy-eight petitions had been presented to Parliament. One of these, from Project Jonah, included 145,000 signatures. The Australian Conservation Foundation presented a petition with 40,000 signatures.

Australia was the last English-speaking country in the world still whaling. The United Kingdom, Canada and New Zealand had stopped whaling for economic reasons. In 1973 the UK had banned the import of whale products, with the exception of sperm whale oil and ambergris. New Zealand banned the importation of whale products in 1975. The United States stopped whaling in 1972.

Graffiti started appearing in Sydney and Melbourne during the 1977 federal election campaign:

Phoebe, I love you
Mr Fraser, please stop killing whales

Some were twenty metres long and done in rainbow colours. Phoebe was popular. The biggest was a large blue whale, on a hoarding at Circular Quay where the Manly ferries berthed, saying:

Thank You, Phoebe!

In Sydney, the eastern suburbs railway works sported graffiti in red paint:

Life wasn't meant to be easy for whales

That one — a play on an aphorism made famous by the prime minister himself — needed two people to get the job done. Richard Jones held Jonny Lewis's legs as the photographer worked upside-down over the edge of the rail bridge.

16 : King Wave : 14 March 1978

A man ran out onto the road as Barbara Cruickshank approached the turnoff to The Gap and Natural Bridge, famous granite rock formations off Frenchman Bay Road. Gusting winds made the car shudder as she slowed. She was heading home from work to one of the cottages at the whaling station, where her then husband worked as radio operator and paymaster.

The man was waving and shouting, and after Barbara calmed him down he told her his friend Stephen had been washed into the sea under the Natural Bridge. One moment Stephen was standing, the next a huge wave came out of the sea and when the water retreated, his friend was gone. He thought Stephen couldn't possibly have survived but, when he peered over the cliff, there he was, trying to swim out to sea away from the churning water around the rocks.

Five minutes, 5.8 kilometres from the turnoff to the whaling station — Barbara did it in half that, her heart thumping. The whaling station was shut for the day. The whale chasers had had no luck and most had gone home early. She found Mick Walters, the station manager. He called John Bell, spotter pilot, and then picked up the handset of the two-way.

Stephen Matthews marvelled at the Natural Bridge, a forty-five-metre arch of granite rock on the coast at Torndirrup National Park, established in 1918, one of Western Australia's oldest national parks. One month off his twenty-first birthday, he took his new camera, an early birthday present, closer for a better picture. The span of the bridge was twenty metres wide and ten metres thick. Twenty metres below, huge waves created a turbulent backwash as they smashed into the granite cliffs.

He'd been travelling with a holidaying couple that drove an old car and towed a bubble caravan behind. They'd picked him up in South Australia, responding to his hitchhiker's thumb.

When they got to Albany they did the round of tourist sights. At the Natural Bridge, the swell was rolling in from the south-west, smashing against the rocks and sending sheets of spray into the air. Stephen's new leather jacket protected him from the cold but didn't dull the sheer force of the landscape. The waves and the wind drowned out all other sound. The smell and feel of the sea was overwhelming. People were visitors, individuals moving warily against the power of the landscape.

The viewing area restricted the view through the arch of the bridge. It looked down on, rather than through, the granite frame of the bridge. The fit and agile could scramble down one side to a flat area level with the base of the Natural Bridge. Stephen carefully picked his way to where he could see straight through the span of the bridge, a giant window to the water beyond. He regretted his decision that morning to slip on a pair of thongs. The rocks where he stood were dry; the waves obviously didn't get this far. There were plenty of other people there too, wanting the better view.

Stephen got out his camera. The next wave broke against the

rocks like all the others before it, but this one didn't recede. It kept coming and coming. Stephen looked up from his camera, turned and ran. The man beside him, closer to the advancing wave, also took off. The wave caught up with them. The black rocks, which moments before had been as dry as desert boulders, turned slippery. The other tourist, well shod, made it to safety but Stephen was caught.

His thongs were the first to go, dragged from his feet by the force of the water. Stephen followed, pulled down the slope by the retreating water, his backside bumping over the rocks. He fell into a void, dropped to churning white water and went under. He bobbed up, grabbed a breath and was sucked under again, where he was thrown around like a rag doll. He didn't know if he was up or down. He came up again some distance from the cliffs.

The light was dimming. Gasping for breath, he could see people waving on the rocks above. He waved back. At least they knew he was still alive. He was sucked under again.

Struggling out of his leather jacket, he started swimming. Where he'd fallen in was like being locked inside a washing machine. He wasn't going there again. He hoped that there could be a way to get to land, or an island, if he kept following the coast. He struck out on a diagonal route, one that allowed for the drift of the current and the power of the swell.

The whale hunters of Albany sometimes had days when no whales were sighted and would return to town early. *Cheynes II* was the last chaser due home that day. Nightfall was thirty minutes away when the radio came to life. A tourist had gone into the water near the Natural Bridge, about seven nautical miles from the chaser's position. It was 6pm.

Paddy Hart ordered full speed from the engines and turned the *Cheynes II* west. He thought about the man who'd been dragged into the ocean. Poor bloke didn't have much chance. The waves crashing against the sheer cliffs were huge. Sometimes a king wave, a combination of waves that grew into a mountain of water, would form, dragging whatever was under the Natural Bridge into the sea. If you were caught in one of these, holding on wouldn't help. The power of the waves could tear anything not bolted down into the sea.

But even if the man had survived being dragged in and had managed to struggle past the breakers, there was no safe place for him to land, no beach or shallow rock shelf to clamber on to. His only chance was to swim away from the rocks and that would take guts and a level head.

Paddy's challenge was to work out how to approach the problem once he got the chaser to the search area. The chances of getting the ship close enough to the cliffs to pick up the tourist, even if he was still alive, were remote. No-one knew, with any accuracy, the depth of the water or where the submerged rocks were; there were no charts for operating that close to shore. It was probably easier to pick the winner at the Albany horse races than spot something as small as a head in the dark in a big swell. How would they find the bloke?

A call came into the Albany Police Station from the ranger at Torndirrup National Park. Sergeant Terry Goodman gathered as much rescue equipment as he had, loaded up and headed out from Stirling Terrace base with three other officers. The time was 6.07 pm.

As they sped along Frenchman Bay Road, the sergeant sent a radio message back to the station asking that the whaling station be contacted for assistance with a plane and any vessels available. The reply came: the *Cheynes II* was already on its way and the spotter plane, piloted by John Bell, would arrive in the area shortly.

The light was fading when the police arrived at the Natural Bridge. Goodman could still see the man in the water, three to four hundred metres away, almost parallel to the cliffs and drifting northwest. There was no set pattern to the waves. The sea was running at three to four metres. The man in the water had managed to get himself outside the impact zone, the area of white water where the waves hit the cliffs and surged back. He was afloat and alive.

The first task was to secure a line of communication. Goodman needed to get as much information as possible to the *Cheynes II*. Time was running out. A rescue would be that much more difficult when nightfall came. He stationed Constable Maria Coyne in a car in the Natural Bridge car park. The limited power of his hand radio at the cliff could reach her, and she would relay messages using the car's more powerful radio to the base at the police station. The two constables at the station would then pass messages to the whaling station, which would in turn pass it to the spotter plane and the *Cheynes II*. Messages from the plane and the whale chaser would come to Goodman back along the same chain.

He and Constable Coyne built a bonfire to act as a beacon for the *Cheynes II*. The spotter plane, with its landing lights on, was making sweeps so low that sometimes the aircraft was below Sergeant Goodman's vantage point on the cliff.

When the call came for help, John Bell had to borrow a plane. There was no time to get his usual aircraft ready. He couldn't see a thing. He had no hope of picking up something as small as a person in that swell and with the light fading. The only thing he could do was send a message to the tourist. He wanted the man, if he was still alive, to know that people were trying to help him. John switched on all the lights on the plane and went as close to sea level as he dared.

Hang on. Don't give up. We're coming for you.

Paddy saw the bonfire on the cliff and ordered the searchlights on. The time was 6.56 pm.

The *Cheynes II* turned to start a circular search. The ship tipped sideways at thirty-five degrees as it came parallel to the waves.

Chatter ... chatter ... chatter ... The squawking radio.

Paddy was annoyed. He had to listen to John in the plane on one radio and switch to another to get messages, via the whaling station, from the police. Chaos ruled. Messages were coming from everywhere: from the station, from the aircraft, the police.

The *Cheynes II* was difficult to manoeuvre at slow speeds and the lack of depth was a constant worry. No-one spoke about it aloud but the ship drew sixteen foot (4.88m) and the depth in front of the cliffs varied from twenty-seven fathoms (49m) to nine fathoms (16m). In the troughs of the waves, the ship's keel bounced closer to the bottom. It was getting darker and Paddy couldn't hear anything but the noise from the radios and the ship's engines. A message from the police, via the whaling station, told him the ship was near to where the tourist was last seen.

Paddy decided to shut down: the radios, the engines, even the people. He wanted complete silence. He ordered the engines to full stop, switched off the radio and told the crew to listen hard. The men strained to see in the darkness, their ears tuned to any noise other than that of the surf.

The ship rolled heavily in the big swell. Sergeant Goodman, watching from the cliff top, saw the hull of the forty-seven-metre *Cheynes II* disappear in a trough between the waves. One minute the ship was riding high, the next it was gone. He counted four circular sweeps before the *Cheynes II* came to a stop.

Below decks, chief engineer Bob Wych knew the ship was close to the rocks. The hull was a sounding board and he could hear the grind of the waves against the cliffs. The swell was rolling in from the ocean, crashing into the cliffs and surging back. The ship was being pushed from two sides. Hidden rock pinnacles lurked near the surface.

Colin Westerberg, related to the local fishing family that helped start the Cheynes Beach Whaling Company, was in the barrel. The twenty-four-year-old had confidence in Paddy, a skipper who ran a disciplined ship. With seventeen blokes, you needed them working together, each doing the task they were given.

Colin saw a dark lump in the water, then picked up a sound,

something not made by the wind or the water. He shouted and pointed in the direction of the object.

The searchlight followed. There was the tourist, his head above the water.

Paddy could tell that the man wouldn't make it on his own. He was stuffed, exhausted. Paddy needed to move the *Cheynes II* closer, making sure the ship didn't run aground. He also didn't want the ship to drift in on top of the bloke and knock him under.

The first attempt failed. Paddy went around for another try. The cliffs were too close, the back surge from the swell too unpredictable. The *Cheynes II* was within fifty metres of the man in the water. That was it. Paddy couldn't risk getting closer.

Keith Richardson, Paddy's first mate, stripped down to his underpants, grabbed a rope, tied it around his waist and dived overboard.

'Keith was the bravest bloke,' says Bob Wych. 'He dived in. No fear of anything. He just grabbed a line and went in.'

Stephen Matthews was slowly losing the struggle against the waves, the swell and current. As a boy in Adelaide, he had spent weekends swimming at the beach or the local pool, but this was different. He had abandoned his plan to swim around the coast or strike out to an island because he knew he risked ending up on the rocks. He had to concentrate on staying afloat and away from the cliffs.

A light aircraft flew overhead. Its lights appeared close.

'Help.'

No-one heard.

Out of the gloom he caught a glimpse of something. A mast? The seas seemed to be getting rougher. In between dips in the swell, the mast turned into a ship. It was closer.

'Help!'

An eternity later, a large man swam out of the darkness and huge arms grabbed him around the chest.

'Am I glad to see you, mate.'

The man didn't respond.

Time accelerated.

As the crew was hauling them in, a wave picked them up and dropped them on the deck. Amidships, the chasers are low to the waterline for easier access to the whales tied alongside. As the wave receded Colin grabbed hold of Keith, over six feet and solid as a wheat silo.

Stephen tried to stand but couldn't. There was something wrong with his leg. He lay on the deck. The shock hit. The shaking and shivering wouldn't stop. The whalers checked him over and found cuts and bruises, and a broken ankle. He had been in the water for around one hour and forty-five minutes.

Keith had cuts and scrapes on his feet, legs and hands from the barnacles on the side of the ship.

The men concentrated on Stephen, the person they had worked so hard to find. A sharp, metallic bang. The crew froze, fearing the hull had hit a submerged rock. A split-second later, they realised that someone must have dropped a steel bucket in the engine room.

They exploded into action. They wanted to get out of there as soon as possible. The time was 7.10 pm.

Sergeant Terry Goodman investigated previous incidents of people swept into the sea off the Natural Bridge by king waves and found that Stephen Matthews was the only recorded survivor. He prepared two submissions that recognised the courage of Paddy Hart, skipper, and Keith Richardson, first mate. He wrote: 'There is no doubt whatsoever that without the actions of Hart and Richardson in this rescue that Matthews would have lost his life.'

The two local government authorities, the Shire of Albany and the Town of Albany, held a reception for the whalers, the police and all involved in the rescue. At the reception, Stephen Matthews was asked if he were a member of Greenpeace. He shook his head.

Keith Richardson and Paddy Hart of the *Cheynes II*.
Photograph by Ed Smidt.

'A pity,' was the response. 'It would've made a good story if you were. Imagine a Greenpeace member being rescued by the whalers and the coppers.'

Stephen spent time in Albany Hospital, then worked on the whale chasers for a few months. He saved enough to buy a motorbike and then took off again on his travels.

Six days later, on 20 March 1978, the re-elected prime minister, Malcolm Fraser, announced the formation of an independent judicial inquiry into whales and whaling.

'Many thousands of Australians — and men, women and children throughout the world — have long felt deep concern about the activities of whalers. There is a natural community disquiet about any activity that threatens the extinction of any animal species. I abhor any such activity — particularly when it is directed against a species as special and intelligent as the whale.

'There are, however, two distinct views in relation to the activities of whalers. One view put to me strongly is that all whale species under threat of extinction are protected by moratoriums imposed by the International Whaling Commission and that current policy is in line with the best principles of conservation. An alternative view, which has also been strongly argued to me, is that the present practice of killing whales does endanger the whale species. Many other arguments also have been put on both sides.'

The fair way to resolve this was an independent inquiry.

Two weeks later, on 4 April 1978, a cyclone struck Albany. No-one had ever heard of a tropical cyclone that far south. The first indication that anything was wrong was when the power went early in the evening and the beer stopped flowing in the pubs. Wind gusts reached 150 kilometres per hour around 11pm, enough force to hold a car, disengaged from the gears, from rolling backward on a reasonable slope. Property, vast tracts of forest, and thousands of livestock were burnt as the cyclone fanned bushfires. Five people died that night in Western Australia, two of them in Princess Royal Harbour as they tried to secure their boat. A woman died up the Albany Highway near Kendenup, when a tree fell on her.

Albany woke on 5 April with a disaster headache. Damage to property ran into the millions. The rural losses would mean fewer farming dollars flowing into town. Industry would suffer and there would be fewer jobs.

Six days after the cyclone, 10 April 1978, the directors of Cheynes Beach Holdings Limited signed off on the 1977 annual report. The 311 shareholders received a final dividend of eighteen cents per share, making a total of twenty-five cents for the year. Profits were $362,906, compared with $539,117 for the previous year. Most of the sperm whale oil production of 4,154 tonnes was sold at good prices. Prices had fallen since then. The oil tonnage was down on the previous year due to a fall in the catch quota.

The Cheynes Beach Whaling Company did well from the Canberra whaling commission meeting in terms of a catch quota for the next season. It was set at 713 whales (536 males and 177 females), eighty-nine more whales than in 1977.

The directors announced: 'The company regards this increase as supporting the view held by the scientific committee of the International Whaling Commission that stocks of sperm whales off Albany are in a stable condition based on scientific assessment and monitoring which has been going on for some fourteen years.'

The long-term demand for crude sperm whale oil had been reaffirmed by a recent market investigation.

'We are assured that for the multitude of purposes for which crude sperm oil is being used there are not satisfactory commercial substitutes available. The major concern expressed by end users is that a steady supply at a stable price for crude sperm oil should be maintained. Price levels have dropped from those achieved for oil sales made last year but all market sources indicated that a long-term stable price at economic levels can be anticipated.'

The directors noted that the Australian Government had established a public inquiry into whaling. 'The company is confident that the findings will support continuation of the company's whaling operation.'

17 : Blood : June 1978

The organisers of the 1978 International Whaling Commission meeting in London were expecting trouble. After the Canberra meeting the year before, the commission tightened the rules for accreditation to make it harder for non-government groups to get inside meetings. Reporters were barred from all except the opening and closing sessions.

Richard Jones flew from Sydney to London. He caught up with Canadian David McTaggart, head of the newly formed Greenpeace international, who would be at the meeting as an official observer. What were they going to do to keep the campaign moving? Richard thought it would be good to take over the meeting. David liked the idea.

Richard briefed the team to say they were conservationists, not Greenpeace members, knowing what he planned would not be supported by the mainstream conservation movement. To complete his private plan — a plan he had shared with only one other person,

David McTaggart — he needed blood. The real thing, not red ink. He went to an abattoir at Feltham, bought two litres of de-coagulated calf's blood and decanted it into two recycled grapefruit juice bottles that could be hidden in the pockets of his anorak.

On the first day of the meeting, he spent some time talking to the policemen standing guard at the front of the Mount Royal Hotel in London's West End. They got to know Richard's face.

He met with the other conservationists at an unlocked rear door to the hotel. When they were sure the press was there, they all filed into the thirtieth meeting of the International Whaling Commission. The faces of those in the conference hall told the story. What the hell was going on?

But the protesters were peaceful. They handed out scrolls alleging crimes against nature to the whaling nations of Iceland, Russia and Japan.

Richard was the last to head for the door when they marched out. He appeared calm and purposeful, but inside he was tense and keyed up. He nodded to David. He walked casually behind the Japanese delegation and slipped his hand into his jacket. He brought out a bottle and poured the bloody contents over the working papers of the Japanese delegation, splashing the Icelanders in the process. The whole place erupted into pandemonium. The delegates thought they had a terrorist.

The Japanese delegation grabbed Richard by his collar. David ran across the room and, assisted by a hotel security guard, prised the fingers from around Richard's throat. The guard was punched in the kidneys for his trouble.

Richard shouted at the Japanese: 'You're an uncivilised nation!'

The Japanese screamed: 'We're not uncivilised!'

Fearing a fistfight, David and the security guard took Richard to another room. When the guard had left, David and Richard took a seat to catch their breath.

A policeman entered, one of those Richard had talked with

earlier. 'What's going on?' he wanted to know.

'Someone poured blood over the Japanese,' Richard said. 'It's all over now, everything's okay.'

The policeman frowned.

David jumped in. 'You know, we're going to take legal action. This is terrible.'

Richard interrupted. 'He doesn't know what he's talking about,' he said, nodding his head in David's direction. 'He doesn't know anything about English law. Don't worry about him.'

Alarms sounded. Richard opened the door to see orange smoke filling the building, obviously from a smoke bomb. He headed to the exit with the rest of the evacuees, expecting with every step to feel a policeman's hand on his shoulder. He was amazed to find himself on the Underground, his girlfriend beside him, travelling away from the whaling commission meeting. He had got clean away.

Richard's action upset the conservative side of the environmental movement. This was not a respectable way to help the whales. This could damage the cause.

Call it a stunt. Call it the work of ratbags. Richard had no doubt about the effectiveness of the incident. It had lifted the profile of the campaign. Normally the anti-whaling lobby had little coverage in the UK press, but the blood incident brought huge international media attention, including newspaper front pages in the UK and television round the world. Richard's brother watched it in his home on the other side of the world in New Zealand.

Catch quotas for whales are set by the International Whaling Commission using a recipe based on sightings, the effort put into catching whales and a string of mathematical formulae. The key is to estimate, with some accuracy, the population of each type of whale ranging over thousands of nautical miles. If the population base is known, it's possible to calculate how many whales can be culled, or killed, to obtain a sustainable yield without reducing the population to a point of collapse. But it's not like counting sheep in a paddock, educated guesses, albeit based on sound theory, are the name of the game.

The key set of data used to calculate sperm whale numbers is collected by spotter aircraft operated by whaling companies. The pilots make estimates from the whales they see.

The scientific committee of the whaling commission, meeting at Cambridge in 1978, considered a proposal to reassess the size of the Division 5 sperm whale population. The paper was titled 'Revised Abundance Indices for Sperm Whales off Albany, Western Australia'. As a result, the committee agreed that a new calculation of the sperm whale population off Albany should exclude data taken before 1967. The small floatplane, as flown by John Bell in his rescue of Ches Stubbs, was replaced that year by more efficient, long-range twin-engine aircraft. Including the earlier data, it could artificially inflate the population estimates to look like there had been a natural increase in the sightings of sperm whales. But, of course, more whales were seen post-1967 because the bigger aircraft covered larger areas of ocean.

However, there wasn't time to do the actual calculations and create a new catch quota. This would have to wait until later in the year.

Sir Sydney Frost, the former judge now heading Australia's independent inquiry into whales and whaling, attended the meeting. He found the informal discussions with the scientists of great assistance. They gave him a deeper understanding of how the whaling commission set their figures.

18 : The Judge and Time of Death : July 1978

Sir Sydney Frost heard the voice on the radio: 'Blowers ahead.'

The hunt, the up-close chase, had begun. Sir Sydney, as head of Australia's Independent Inquiry into Whales and Whaling, was determined to see, firsthand, the operations of Australia's three remaining whale chasers. He'd been out with the chasers before but the whalers had been unable to catch a single whale. He wondered if his presence on board had spoiled the aim of the harpoon gunner.

Sir Sydney had left his bed at Albany's Dog Rock Motel when it was still dark to join skipper Gordon Cruickshank and the crew of the *Cheynes III* on their daily trip to the edge of the continental shelf.

Now it was 1.55 pm and it had been a long and painful journey from Princess Royal Harbour. The weather was fair, by whalers' reckoning, but Sir Sydney was seriously seasick. Most of his day had been spent lying on the deck. The crew checked on him regularly as he was rolling around. He could have rolled right off the boat.

It was a long way from tropical Papua New Guinea, where Sir

Sydney was that country's first chief justice at independence in 1975, or from Melbourne, where he had grown up and now lived in retirement. Despite his sickness, Sir Sydney watched and took detailed notes. He studied the tactics of the whale chaser, the behaviour of the whales, the duties of the crew, what was said, the equipment used and the weather. He asked a lot of questions and spent a long time talking to sonar operator Tom Kennedy. Tom could see that the judge didn't like being at sea.

Mick Walters, operations manager and spotter pilot, had spent eleven years honing his skills since he had joined the whaling company in July 1967. So far that year, Mick had recorded six thousand whale sightings, most of them sperm whales. He'd recorded sightings of thirty of the giant blue whales, plus fin whales, sei whales and hundreds of killer whales. Blue whales are the largest animals in the world. The largest ever caught was over thirty-four metres and weighed about 174 tonnes.

Mick flew the plane one-handed while he operated a camera with the other. He had a collection of photographs showing pods of two to three hundred individual whales. Each season, lasting 160 to 180 operational days, the spotter plane would be in the air 1,700 to 1,800 hours. This was six to seven hundred hours more than the prior decade. He was convinced the numbers of sperm whales hadn't changed. He knew there were a lot of whales out there and he was sure they were in no danger of extinction.

On the *Cheynes III* the radio came alive. Mick Walters reported more whales five kilometres ahead and the *Cheynes III* took up the chase. Skipper Gordon Cruickshank picked on one pair as the pod split, both males on the big side, around thirteen and fourteen metres in length.

Sir Sydney saw the creatures go into the shallow dive pattern — submerge and surface, submerge and surface — at intervals of seven

Whale chasers, Australia's last whaling fleet.
Photograph by Ed Smidt.

to ten seconds. Gordon ran along the walkway from the bridge to the gun deck. A sharp cracking sound as the cannon went off. The harpoon flew fifteen metres off the port bow and hit the sperm whale behind the flipper and forward of the animal's centre.

Sir Sydney recorded the time as 3.38 pm. Gordon told the judge he was pretty sure it was a clean kill. He told the deckhands to prepare a killer harpoon to make sure.

Sir Sydney timed the deckhands. It took two minutes to load the killer, the wooden harpoon with an explosive head, no flukes.

The chaser was still travelling. Gordon had to move the ship to make it easier to bring the whale alongside. He shouted to the bridge, 'Go astern. Go astern.'

The crew started the winch and brought the whale close. A killer shot was fired.

Sir Sydney timed the procedure at five minutes. He concluded it would take longer if conditions weren't as good.

Gordon agreed. 'It could take a little bit longer but not all that much longer.'

The whale, now on the starboard side, was pumped full of air. The tail was trimmed, a hole cut in it and a rope attached with marker buoy and radio beacon.

The *Cheynes III* turned to pursue the second whale. The spotter plane was busy directing one of the two other ships and couldn't help Gordon and his crew, but sonar operator Tom Kennedy picked up the trail without too much difficulty.

According to Sir Sydney's watch, the second whale was caught at 4.22 pm.

Gordon got in close and shot it on its right-hand side. 'I'll take that alongside on the starboard side,' he told Sir Sydney. 'That one was instantaneous. That one is dead.'

Sir Sydney leaned out to get a better look. The whale vomited a piece of squid into the sea next to the ship and a big chunk, with tentacles attached, floated free. Sir Sydney estimated it to be as thick as a man's body and 2.5 metres long. The crew talked about sperm whales with squid sucker marks as large as dinner plates.

One of the terms of reference for the inquiry was to examine the methods of killing whales and ascertain whether or not it was humane. Gordon told Sir Sydney that in most cases death was quick. A killer shot was used to make sure. He explained his definition of instantaneous death. 'If the whale comes up and turns on his side and his mouth is open, I take that to be instantaneous. If he comes up and thrashes a bit, we bring him alongside to put a killer in him.'

'How much time do you think would normally be required between the first harpoon and the killer harpoon?' Sir Sydney asked.

'With everything going right, it would be three minutes because you're heaving in at the same time as you're loading the killer harpoon.'

'And if a second or third killer was required, how much time might there be between?'

'You wouldn't have to heave in. You'd only need to load the killer.'

'Would it be down to two minutes?'

'A couple of minutes, yes.'

With the commissioner on board, Gordon didn't want to be too late back to base. He called it a day, picked up the first whale and headed home, a whale tied to each side of the *Cheynes III*.

Sir Sydney checked his watch again. 'How long do you think it will be before we're back?'

'I'll have to work it out,' Gordon said. They were seventy kilometres east of Albany. If all went well, they would be in harbour at 9.15 pm.

The *Cheynes III* hit a current. They arrived at 10.30 pm.

Later, Sir Sydney followed up with Mick Walters about harpoons and whether the whales suffer when caught.

As operations manager, it was Mick's job to ensure the whales were caught and killed as quickly as possible. There was a sound commercial reason for this. A quick turnaround meant a whale chaser had a better chance of catching second and subsequent whales. He told Sir Sydney that the time between the first harpoon and the killer shot was largely academic. The first harpoon was the effective one.

'When that enters the whale's body and the grenade explodes, it's hitting with the force of a 135-pound harpoon and the twenty-three-pound grenade at the end of it, a force greater than a twenty-five-pound shell. Any ex-serviceman who has had the misfortune to be wounded will tell you there is a numbing effect after a gunshot wound, whether it's explosive or bullet, and that effect remains for some time. Although the whale may be thrashing around, it's still in a state of numbed shock without any great deal of pain. The second shot, which is put in at point blank range, invariably will, if the whale's not already dead but is still in a state of numbed shock, kill him.'

19 : The Inquiry : 31 July 1978

Tony Gregory from Sydney and Stephen Kaye from Melbourne flew to Perth, hired a car and drove to Albany. The Project Jonah pair did the road trip in less than four hours, booked into the Dog Rock Motel, freshened up and went to the restaurant. Sir Sydney Frost was in the dining room with John Saleeba of the whaling company. It was the night before the start of the inquiry.

Dr Peter 'Bro' Brotherton had been preparing for this day for a long time. After long years of campaigning, it came down to convincing one man, a judge, that whaling wasn't something in which Australia should be involved. Bro was keyed up. He had one shot at making whaling disappear.

Project Jonah had $25,000 in funding from the federal government to help pay its legal expenses at the inquiry. But Bro, Friends of the Earth representative for Western Australia, had no funding. He came down by car and stayed at a backpacker hostel out of town.

Tara Hall, a room for hire above a Chinese restaurant in the centre of Albany, had thirty people waiting for the inquiry's first day of formal hearings to start. A table at the front was populated by a who's who of the anti-whaling lobby: Tony Gregory and Bob McMillan of Project Jonah in New South Wales, Barbara Hutton of Friends of the Earth in Victoria, and Bro Brotherton from Friends of the Earth in Western Australia. They had all applied to be represented at the inquiry and were serious and busy, checking files, making sure they had everything they needed.

Another table contained managers from the Cheynes Beach Whaling Company, including executive director John Saleeba. Nearby sat state and federal government representatives, including Ken Marshall, the region's senior Western Australian government officer.

Technicians, seconded to the inquiry from Hansard at Parliament in Canberra, put the sound system together, the polished wooden floor becoming a mass of electrical cable.

Sir Sydney had already received written submissions from 102 organisations and seventy-three individuals. He read each of them and made handwritten notes. The submissions came from all over the world: from conservation groups, from the whaling industry in Japan, and from governments. Albany residents had made submissions as well, not all of them pro-whaling. Sir Sydney had also commissioned specialist reports: a review of current whale harvesting; and whale brains and their meaning for intelligence. The judge had travelled and met with twenty-eight scientific experts from North America, the United Kingdom and Europe.

The secretary of the inquiry, Andrew Struik, a bearded public servant from Canberra, spoke: 'This is the first public hearing of the inquiry into whales and whaling which, as you know, is being conducted by Sir Sydney Frost. This hearing will now open.'

The first session of the inquiry was scheduled to continue until lunch but it would last only twenty minutes.

Sir Sydney, wearing a dark suit, a stark-white shirt and equally bleached handkerchief in his pocket, clasped his pale hands in front of him. 'Ladies and gentlemen, I have been informed by Mr Saleeba ... of Cheynes Beach Holdings, that he has an important announcement which must affect the course of this inquiry.'

John Saleeba, the public spokesman for the whaling industry, left his seat to take the witness table. He had a sheet of paper in his hand. The noise level in the hall rose as people speculated with their neighbours about the significance of this departure from the inquiry schedule.

An image of the people sitting in Tara Hall burned into John's mind. A report to the shareholders of Cheynes Beach Holdings was at that moment being delivered to the stock exchange, the same information that the company had given to staff that morning at 6am.

The directors of the whaling company were concerned about the profitability of whaling in Albany both for the current year and for the future, John said. As he read from a prepared statement the text of the report to the company shareholders, the audience looked stunned.

'Our licence to operate in 1979 must remain in doubt until the report from the Honourable Sir Sydney Frost on the Inquiry into Whales and Whaling in Australia, due for submission to the federal government on September 30 this year, is considered and a decision made thereon by that government.

'Australia supplies a substantial proportion of the free world market demand for sperm oil, as Russia and Japan largely use their own production domestically. Your company, throughout its history, has sold the major proportion of its oil production as crude sperm oil in bulk to one or more of a limited number of refineries in Europe including the United Kingdom. These refiners in turn sell to numerous end users.

'In the early part of this year our normal buyers showed extreme reluctance to commit themselves forward and this was causing your board concern. But it was not until mid-July following a visit to Europe and the United Kingdom by two of your directors, that it was realised how serious was the move from filtered sperm whale oil to alternatives by the end users. Invariably the reason given was doubt of continuity of supply from Australia. In our opinion this was an unforseen effect of the public inquiry of which the first term of reference is whether Australian whaling should continue or cease.

'Historically the sperm oil market has always been of a volatile nature and it is difficult to foresee. Your directors believe that whaling operations for 1978 will result in a substantial loss. Assuming we receive a whaling licence from the Australian Government for 1979, our quota for that year compared with 1978 has been drastically cut — males by twenty-five per cent and the less productive female by ten per cent — resulting in an estimated drop in production of crude sperm oil of 1,000 tonnes. Although certain direct operating costs should fall in line with the reduced production, it is not possible to cut most overhead charges proportionally and thus total costs per tonne of oil must rise very substantially. Oil revenue comprises nearly eighty per cent of our total whale product sales and it therefore does not seem feasible to forecast any likelihood of profitable whaling operation in 1979.

'In view of the foregoing and in the best interests of shareholders, your directors have decided that whaling operations must cease in the near future. No decision has yet been made on the closure date. Consideration has to be given to forward commitments for meal and solubles, and the needs of the Australian industry for refined oil hitherto supplied by your company.

'There is also the concern of dealing fairly, within the capability of the company, with employees of long standing who may become redundant. After a quarter of a century of whaling with good returns to shareholders, the directors' decision has been made with deep regret and reluctance.'

John stopped reading and looked up at Sir Sydney. 'Sir, I think that is all that is required by the inquiry.'

'Well, Mr Saleeba, has this report already been sent to shareholders?'

'Yes, sir. It should be in the hands of the stock exchange by now also.'

'Yes, very well.' Sir Sydney's voice carried a trace of annoyance. 'Now then, you have said that the directors have decided that whaling must cease in the near future. Would you care to amplify that?'

'Well, sir, I find it very difficult to amplify that. We, as you appreciate, have a responsibility to our employees and a responsibility to our shareholders. It is our desire to extend our operations to provide employees with as much time to adjust as is possible and the actual date of cessation of whaling operations will depend on inquiries that are currently being undertaken, but the results of which we do not anticipate will be with us for perhaps a week. I am afraid, at the moment, that is as close as I can get, except to say it could be very soon or it may be later this season.'

'In any event, the closure will take place no later than the end of this season?' said Sir Sydney.

'That is correct, sir. The comments which we made in relation to the likely quota for next season, I think, cover the problems we face there.'

'No later than the end of this season and it is possible it will be earlier?'

'Yes, sir.'

'How much earlier depends upon these marketing inquiries which are now taking place?'

'That is correct, sir.'

Sir Sydney turned his attention from John to the centre of the hall and adjourned the inquiry until the afternoon.

Outside, John was questioned by the press. Because his statement to the inquiry did not give a date for closure of the whaling station, he was asked to be unequivocal. He was. The whaling station would close. There was no doubt.

Back at the whaling station, he took calls from reporters around the world. One was from Canada, the birthplace of Greenpeace, asking for confirmation of the announcement. This was world news? he thought. They want to know in Canada?

The global perspective influenced the company's decision to close. Most of the Albany production of sperm whale oil was sold outside Australia. As the free world's biggest producer of sperm whale oil, the Cheynes Beach Whaling Company's output had a large impact on price. In mid-1976, when the whale catch quota was cut forty-five percent, the price jumped to GBP320, up from GBP175 the year before. South Africa pulled out of whaling in response to lower catch quotas, further reducing supply. The price of sperm whale oil hit GBP470 per tonne in the third quarter of 1977 and for a time was more than GBP500.

By the first quarter of 1978, the price was in freefall. In the second quarter of that year, the price was down to GBP250 per tonne. The traditional buyers, those who further refined sperm whale oil and the ultimate end users, had lost confidence in a continuing supply and were looking at synthetic alternatives. They had read about the protests at Albany in mid-1977, how Greenpeace had blockaded a shipment of sperm whale oil in North America, and the announcement of the independent inquiry in March 1978. In Europe, this last factor came into sharp focus because the buyers were contacted to provide information to the Australian inquiry. Added to this was the continual revision downwards of the size of the sperm whale population. Quotas, as set by the International Whaling Commission, were falling.

The only way for Cheynes Beach Whaling Commission to maintain profitability was if there were a substantial rise in the price of sperm whale oil, but a big rise would make substitutes more attractive and push sperm whale prices down again.

Tony Gregory of Project Jonah couldn't believe what he'd heard in the inquiry. What's this trick all about? he thought. He didn't know what to do. Should they go home or were they being dudded?

Bro Brotherton was also suspicious. He'd invested a lot of energy in preparing for the inquiry and now, without a fight, the whalers appeared to have conceded defeat. He kept analysing John's statement for loopholes. Was there a way that the whaling company could continue whaling despite what was said to the inquiry? He believed the closure announcement was timed to imply that the inquiry was partly responsible.

He put his thoughts to paper and issued a quick press statement during the adjournment: '… the only definitely established link between the inquiry and the company's impending closure is the deliberate coincidence of the announcement at the beginning of the inquiry. This coincidence does not demonstrate that the inquiry was in any way responsible for the company's decision. The inquiry has merely been used as a convenient scapegoat by the Cheynes Beach Whaling Company.'

With unexpected time on his hands, he gathered some of his conservationist colleagues and headed for the whaling station. 'Five of us decided we would actually, for the first time, go out to Frenchman Bay when there were whales there,' says Bro. 'I thought it was all coming to an end and I probably should go there.'

There was a crowd at the whaling station. Bro had a weak stomach at the best of times, and the smell at the whaling station reminded him of rancid butter mixed with burning car tyres. One end of a whale was on the deck and the other end was in the water. There was a massive saw that they used to cut off the heads. The buildings and equipment were ramshackle and looked like some sort of Dickensian workhouse nightmare. The station was open to all.

The flensing deck at the whaling station.
Photograph by Ed Smidt.

A bone saw severs the head of a whale.
Photograph by Ed Smidt.

On the flensing deck.
Photograph by Ed Smidt.

Greg Trouchet flensing a sperm whale.
Photograph by Ed Smidt.

Strips of blubber.
Photograph by Ed Smidt.

Members of the public wandered in and out without check. Bro thought it should have been closed on health and safety grounds, and public liability.

The sweet harmony of the song 'The Carnival is Over' drifted from the public address system at the whaling station as the Seekers sang their 1965 hit about broken hearts and last goodbyes. Daryl Adams paused from his work with the flensing knife. Earlier in the year he had joined a contracting crew that cut up the sperm whales and moved the pieces into cookers to start the process of separating the oil. The work was dirty: all blood and guts. He wore the oldest shit-clothes he could find. The smell was horrible. The workers called it 'the smell of money'. You got used to that and a sense of humour helped. The whales were hard and slippery on the outside. The head, where the mother lode of oil was found, was like rock. Daryl learnt to get around the flensing deck making good use of the spikes screwed into the soles of his rubber boots. His leg still pained him from when he stepped into the opening to the cooker, where a boiling stew of whale meat starts the process of extracting oil. His leg was put in water immediately but he spent weeks of agony recovering in hospital.

Early that morning, he and his workmates at the whaling station had been told about their last goodbyes. The bastards, he thought. They've never pumped music through the public address system before. He looked up at the railing where tourists stood to watch the whales being cut up. There was a bigger crowd than usual.

John Bell was in town, picking up his car from the garage, when he heard the news.

The mechanic said, 'What are you going to do for a job now the whaling company's finished?'

John was taken aback. He had had no warning that the whaling company had decided to toss in the towel. The staff had not been consulted. John and his wife had opened a small museum at the whaling station. He felt sure that if there'd been a whisper of anything going on he would have been the first to hear.

It was a secret well kept by the company directors. Nobody had a clue.

Bro Brotherton, on his way to Tara Hall for the resumption of the inquiry, picked up a copy of the local newspaper printed on yellow newsprint. The *Albany Advertiser* decided the news was big enough to create a special edition. The last time a non-scheduled newspaper came out in Albany was the abdication of King Edward VIII.

The inquiry resumed at 1.30 pm. The first day had been marked for the hearing of submissions on the economic impact of whaling: employment, tourism, the service industry, and the direct and indirect impact on the economy.

Ken Marshall, the regional administrator representing the WA Government, thought it unlikely there would be whaling again in Albany. 'In the light of the company's announcement, there would seem to be little other than academic discussion which could take place,' he told Sir Sydney. He urged special action by the Commonwealth Government in Canberra to help the local community. 'It would seem that perhaps the company's decision was influenced to some extent by the creation of the inquiry by the government. It seems appropriate that the government should consider quite urgently some actions which would permit the re-employment of the whaling industry employees.'

Reliability of supply was also a key issue. Ritsche Matla, the president of the Albany Chamber of Commerce, said the government was largely responsible for undermining the confidence of sperm whale buyers overseas. 'Some special compensation from the government is urgently required,' she said.

No government representative at the inquiry could commit the

Commonwealth Government to giving special compensation to the employees of Cheynes Beach Whaling Company. A representative of the Commonwealth Employment Service said the full facilities of the local branch were available to those retrenched. This included training schemes and relocation assistance.

Barbara Hutton of Friends of the Earth, Victoria, said it was important to consider the fate of the people who would soon be unemployed. No-one had made a submission to the inquiry on behalf of these people or to find out what form of assistance they needed. 'The government is not directly responsible for closing the station,' she said, 'but it seems that the inquiry did have something to do with it.'

Sir Sydney said he had asked the whaling company if someone could come to the inquiry to talk on behalf of the employees. He did not know if action had been taken.

Ron Heberle, of the South Coast Licensed Fishermen's Association, was shocked when he heard about the closure. He had not been invited to speak at the inquiry but he turned up anyway. He was the inaugural president of the Albany Conservation Society and had been a member until recently when he resigned over the society's anti-whaling stand. Sir Sydney allowed him to speak.

He looked at the conservation lobby representatives sitting at the inquiry. He told them their satisfied faces indicated they thought they had achieved victory, but it was a hollow one. It was likely that the sperm whale quota not taken up by Australia would be used by another whaling nation. 'We've heard a lot of woolly issues on standards of morality,' he said. 'To me, the greatest tragedy in morality has been that people from all over Australia, who consider themselves moralists, felt justified in attempting to destroy the livelihoods of a hundred people in the town of Albany. I believe this is the only moral issue.'

20 : A better way to die? : Tuesday 1 August 1978

Gordon Cruickshank felt like he was on trial. He was being questioned in detail about how he went about catching whales. He had swapped the grey overalls he liked to wear as the master of the *Cheynes III* for a shirt, tie and jacket when he presented himself at the inquiry at 9.30 am.

The inquiry was pursuing one of its terms of reference: methods of killing whales. Project Jonah had done a great deal of work to calculate how many harpoons it took to kill a whale and how long it took for the creature to die. Bob McMillan, the checker, had become an expert in the area, drawing research from around the world. He'd worked out that it took, on average, 1.8 harpoons.

Melbourne barrister Stephen Kaye was questioning Gordon on behalf of Project Jonah. He asked Gordon if he was concerned about the way whales were killed and, if there was a better way to do it, would he prefer to do the job that way?

Gordon said, 'Is there a better manner of getting killed?'

The whaling company called a meeting of employees for 2 August at the community centre hall to explain in more detail what was going to happen. How long could operations continue? How many would lose their jobs immediately? And how many could stay until the end?

Among those listening in the hall was Joe Isherwood, an industrial officer with the Australian Workers Union. He took detailed notes.

The workers endorsed the view of the company, that the federal government would most likely be delighted to see the whaling station close down. They all thought federal politicians couldn't care less.

The majority of workers at the Cheynes Beach Whaling Company had no chance of a similar job. The engineers, those with expertise and qualifications with steam engines, would need to retrain, even if they could find a job. Those wanting to stay connected with the sea would have to find places within the fishing industry or move somewhere else to take up work with trawlers. This would be a marked change from the daily trips to sea of whalers, as trawlers might operate at sea for up to fifty days at a time.

Albany's work force in employment was about five thousand, spread across a number of mainly service industries including government, retailing and some manufacturing. There were more than seven hundred unemployed already. There were 1,255 people employed by the seven major industries, including the Cheynes Beach Whaling Company: the abattoirs Borthwick & Sons, Hunts Canning Factory, Southern Ocean Fish Processing, Albany Woollen Mills, Albany Wool Stores, Albany Superphosphate Company.

The stability and growth of Albany was tied to the health of the Port of Albany. The annual value of whale oil shipped through the port was about $2.1 million: 3,535 tonnes to the UK, 1,600 tonnes to the Netherlands, fifty tonnes to Japan. The fuel oil sold to the whaling company — around 4,500 tonnes per year — was twenty-one per cent of the total bunker trade through the port.

Wages paid by the whaling company totalled $1.17 million

Cheynes Beach Whaling Company announces closure of Australia's last whaling station. Industry workers meet.
Photograph by Ed Smidt.

spread across 102 workers. This money supported 199 dependents, wives and children. Thirty-one were aged over forty-five years old.

For the rest of the afternoon following the meeting of whaling company employees, two union officials visited local employers. At the Main Roads Department, there was no prospect of taking on ex-whaling workers. The department had budget problems. The Public Works Department: no vacancies. Borthwick's, the meat-processing factory: no vacancies. Cooperative Bulk Handling (CBH): no vacancies. Southern Ocean Fish Processors and Trawling: definitely no vacancies. Albany had one of the highest unemployment rates in the state of around six per cent.

Joe Isherwood could tell it wasn't an election year. This was a serious crisis for these men and for their town, and yet the federal government was silent. He wrote: 'Albany is being left for dead as if it doesn't exist.'

Sir Sydney cut short the Albany hearings. He had been due to hear submissions about the economic impact of whaling on the town, but the closure announcement had made that line of inquiry redundant. Hearings went ahead in Perth, Sydney and Melbourne.

In Melbourne, the funding given to Project Jonah by the federal government to help its representation at the inquiry had dried up. They didn't have enough money to hire a barrister, so Tony Gregory, the New South Wales head of Project Jonah, did the job.

The Project Jonah team produced a flow of witnesses. Three sisters, who ran a group called The Ark, a Catholic animal welfare association, gave evidence in black dresses, black hats, black gloves and black bags. Respectable. Not at all like Greenies, who were generally suspected of being from the ratbag side of society.

Sometimes the value of the help was hard to assess. One man wanted to give evidence about his sexual relationship with a dolphin.

Laurie Levy, a cameraman for Channel Nine, acted as informal media liaison officer for Project Jonah. His work had shown him, firsthand, how the media could shape public opinion. The amazing work of Canadian Bob Hunter was a good example. Laurie was inspired by the images of people putting themselves between the whales and the harpoons, which had put the issue of whaling into the lounge rooms of the world.

Laurie's job was to win media coverage of the people who gave evidence at the inquiry. He wanted to establish a strong connection, through the media, with the public. Australians had always liked the underdog, and whales were the ultimate underdog. People had the resources to fight. Whales didn't. Someone had to speak for the whales.

Dr Michael Greenwood, an American expert on marine mammals and their use in war, had worked for a string of government agencies, including the US Department of the Navy and as scientific adviser to the CIA (Central Intelligence Agency). He agreed to travel to Australia to give evidence on the proviso that he not be compelled to answer questions relating to the US agencies' operational matters or reveal classified information. Sir Sydney saw no reason for the inquiry to involve itself in state secrets. His interest was in the special features of whales and dolphins.

Dr Greenwood had worked with dolphins in and outside captivity. He thought the inquiry was an intelligent way of resolving a problem. He told Sir Sydney, 'I would suggest to you that the gentle sound of your pen as you sign your final recommendation will, in the truer sense, be a sound that will be heard around the world.'

Many had put the argument that dolphins and whales can't be that smart because they allowed themselves to be caught or harpooned, and they kept returning to the area where man hunted. Dr Greenwood argued that the same could be applied to man. For example, there were many reports of people running *into* a fire rather than *away* from danger. The disadvantage for whales and dolphins might be that they had been naive enough to trust man.

The US Navy was first attracted to whales and dolphins, Dr Greenwood explained, because of their hydrodynamic design. What they were trying to do with submarines was mimic the way sea mammals operated. If they could learn from dolphins, some of their design characteristics could be incorporated into submarines.

Another key area of interest was the use of sonar. Dolphins and whales used sounds to locate and identify objects some distance away. There was good evidence that dolphins and whales could determine what the object was: a fish, another mammal, or even a human. A dolphin saw a three-dimensional image acoustically in a 180-degree arc from its brain casing.

Dr Greenwood became involved in this research during the sea lab program, where divers were put down to depths of up to 420 metres and lived on the sea floor for weeks. These were the Homo Aquaticus foreshadowed by the French underwater explorer Jacques Cousteau. Dolphins, unencumbered by the need to make lengthy decompression stops on the way to the surface, were used to carry loads to the sea floor and back again.

He told the inquiry about Dolly, a lead dolphin he released to the open ocean off Key West in Florida in 1969 when the project was being moved to Hawaii. Dr Greenwood thought Dolly had contributed enough; upon release she swam away with enthusiasm.

A man who had lost his job when the Florida program closed had moved about three hundred kilometres up the Keys to live. Six months after the program closed he went down to the beach in front of his house to find Dolly waiting for him. In some inexplicable way she had found him. She stayed until she died in 1974.

'What is important in this anecdote of the animal's behaviour,' Dr Greenwood told Sir Sydney, 'is the concept of memory and the concept of very sophisticated behaviour that had been trained into her, literally, years before, specifically in 1966.

'Adhering to the restraints that I impose on myself, I can nonetheless talk of the initial use of Dolly when she, Dolly, with three other dolphins, was deployed to help locate and recover a nuclear warhead that was lost through misadventure off Puerto Rico, a project known as Green River. The animals did not, in fact, locate the lost device but so convincingly demonstrated that, if they had been given an extra hour or so, they clearly would have located it.'

It was this demonstration in 1966 that led to the development of weapons systems employing dolphins.

Dr Greenwood was asked about reports of dolphins being used during the Vietnam War to bomb a harbour.

'I would prefer not to comment on that,' he said.

21 : The last whale : 21 November 1978

The whale chasers dressed up for their last day. Each of the three vessels flew flags and bunting. The crews were resigned; the industry was gone and their jobs as well. Today they were determined to have some fun. They tried for a carnival atmosphere.

Those who had never fired the harpoon cannon got their chance. Skipper Paddy Hart thought he'd give everyone a go so they would recall, years later, that they'd actually sent a harpoon flying. 'We discharged a lot of shells,' he says.

The mayor of Albany, Harold Smith, who had fought the political battle on behalf of the whalers, and Ken Marshall, the district's senior state public servant, joined the crew on the *Cheynes II*.

'We didn't see a bloody whale all day,' Harold Smith says.

The Cheynes Beach Whaling Company caught nine whales the previous day (20 November 1978). The company's production book records the last whale caught as a female measuring thirty-six feet and three inches, harpooned by Axel (Chris) Christensen, master of the *Cheynes IV*.

Australia's last whaling fleet.
Photograph by Ed Smidt.

Mayor Smith: 'It was a beautiful trip, a lovely day and we had a nice meal, a good steak. It was a great experience. I'd never been before and after that they closed up. They were a good bunch of guys, that's for sure. They accepted the situation and that was the end. It was a good atmosphere. As we came in, the station manager Geoff Reilly fired a harpoon in the direction of the whaling station as a gesture. They came in right close to shore at Frenchman Bay and fired.'

Steam whistles blew as they returned to harbour. Bob Wych, chief engineer from the *Cheynes IV*, pushed the ship hard. He belted the shit out of the engine coming home and they gave a couple of toots coming through the heads.

On the *Cheynes III*, Gordon Cruickshank relaxed the no-alcohol rule and broke out the beers. 'What are they going to do, sack us?' he told the crew.

The souveniring started when they got to the Town Jetty. Most crew got something. Gordon claimed the ship's bell.

There was no special payment or compensation paid to the staff. Mick Stubbs, first mate on the *Cheynes III*, got what he had accumulated in holiday pay plus an end-of-season bonus. As he was within a month of ten years' service he asked that the month be waived. The company agreed. This was the second time Mick had lost a job through closure of whaling. The first time was in the north-west when humpback whaling stopped in the early 1960s. After the Albany closure Mick went fishing and did a variety of jobs. Nothing lasted.

The government made an offer of jobs at Collie in the south-west working the coal mines. No-one took it up.

Bob Wych walked off the boat and went up to the pub with the boys. The following Monday he started work at Cooperative Bulk Handling.

The whaling company sold its stocks of sperm whale oil at GBP260 per tonne, almost half the price it got the year before.

Sir Sydney Frost's recommendations lost their edge in Albany. The whaling station had closed. The jobs were gone. The implications were more international. What would be Australia's position at multilateral groupings such as the International Whaling Commission?

The Frost Report was presented to Prime Minister Malcolm Fraser on 1 December 1978. Sir Sydney gave weight to public opinion. Reasonable Australians would think it wrong to kill an animal of such special significance as a whale. There was no essential human need fulfilled by catching them. The judge agreed with the Project Jonah submission that the continuation of whaling would outrage a significant portion of the population.

Sir Sydney drew on philosopher Peter Singer's submission on ethics: whaling was not morally justifiable if you accepted the position that animals should not be killed or made to suffer pain except when there was no other way of satisfying an important human need. Harpooning whales was inhumane.

Australia should pursue a policy of international opposition to whaling and seek worldwide protection for whales. Whaling should be prohibited within Australia's two-hundred-mile fishing zone and the importation of whale products, or products containing whale products, banned.

Sir Sydney also recommended that current laws regarding whales be replaced by an Act of Parliament to protect them.

The report rejected compensation for the Cheynes Beach Whaling Company and its workers. The company's decision to close before the end of the inquiry ensured that. If the government had forced the closure, compensation would have to have been paid, but the whaling station would have closed no matter what the result of the inquiry. The judge considered that the market trends that led to the closure were firmly in place in the second half of 1977 before the inquiry was announced or established. Scientists at the International Whaling Commission had reached the conclusion that sperm whales off Albany should be protected. The Cheynes Beach Whaling Company would most likely have not been given a catch quota for 1979.

When Canada banned commercial whaling in 1972, three small shore-based whaling stations were still operating, yet the Canadian Government accepted no liability for compensation but made ex gratia payments because the ban may have threatened the existence of small fishing communities.

According to Sir Sydney: 'In the Australian case, to make ex gratia payments to Cheynes Beach and its employees would be to single out the whaling industry for preferential treatment, when a number of Australian companies have had to close down because of changed economic conditions in recent years.'

'Whales and Whaling', a two-volume report, was tabled in Federal Parliament, 20 February 1979. Chris Puplick, the former Young Liberal president who worked with Project Jonah, was now a senator. He moved a motion urging the government to accept the Frost Report's recommendation in full, motion seconded by Alan Missen, the Liberal senator from Victoria who had won the $25,000 government funding for Project Jonah.

Chris Puplick told the Senate: 'If whales some day are gone, also will be gone the opportunity to learn how these warm-blooded social animals survive in cold seas, communicate with song and navigate across thousands of miles of ocean. Gone, too, will be the poetry, the symbol that whales provide us. By respecting their right to live and co-exist peacefully with us, there is evidence that we can find harmony with our environment, with our fellow creatures and, perhaps, with ourselves.'

Malcolm Fraser announced on 4 April 1979 that the Frost Report recommendations would be adopted and the *Whaling Act 1960*, regulating whaling activities, would be repealed and replaced.

Australia has since been a global advocate for an end to whaling.

Returning to port.
Photograph by Ed Smidt.

22 : What happened to them? : 2007

The Whalers

John Bell had done about twelve-thousand-flying hours spotting whales when whaling closed in 1978. He and his wife continued with the small museum they started in a shed at the whaling station, which later became the museum Whale World. John died in a plane crash in 1996 while doing coastal surveillance out of Albany for the Australian Federal Police. He was sixty-two.

    Gordon Cruickshank, master and gunner, became a commercial fisherman. He always said he would go whaling again at the drop of a hat. All he needed was a ship. He died in 2006, aged eighty.

    Paddy (James Frederick) Hart, master and gunner, worked at the Albany Woollen Mills. His pay packet was half as thick as it was when he was whaling but he wouldn't want to see whaling again. 'They were beautiful creatures,' Paddy says. 'Back in them days it was a matter of survival. You didn't think about it. On reflection, it had to come to an end. Why use them if you don't have to?'

Stephen Matthews, who was rescued by Keith Richardson, Paddy Hart and the crew of the *Cheynes II*, and worked briefly on the whalers afterwards, now lives in Adelaide and works as a postman. 'Afterwards, I was a little overcome by it all,' Stephen says. 'The emotion came out. I probably didn't thank enough people when I was there.'

Ches Stubbs, Ahab of the Southern Ocean, died in 1991 aged seventy-five. 'There was a lot of blood and gore,' Ches says of whaling, 'but three to four months along, when you've caught twenty whales or so, you lose feeling. They're like a rabbit or anything else. A pity, but that's the way it is.' His friend John Bell, the pilot who saved his life with a risky landing and take-off at sea, said at the funeral, 'I often wondered if he read *Moby Dick* and decided to live that way.'

Mick Stubbs, first mate and son of Ches, retired in 2006. His last job was at the whaling station, Whale World. His retirement gift was a sperm whale tooth carved in his image. Mick lives in a cottage, itself something of a whaling museum, overlooking the Town Jetty where the whale chasers tied up and the crew ran to make the pub before closing time.

Keith Richardson was awarded a Bravery Medal under the Order of Australia for his role in rescuing Stephen Matthews when he was swept into the sea by a king wave.

Kase (Cees) Van Der Gaag, master and gunner, left whaling on 7 October 1977 soon after the protests in Albany. He worked in the north-west on tugboats and retired to a smallholding off the Frenchman Bay Road on the way to the whaling station.

Kase says, 'I had to defend the thing you can't defend. Killing whales is not something you can defend.'

He and Paddy Hart are generous with their time when it comes to promoting Albany's whaling history. They turn out for media events at the whaling station. They both speak against whaling.

The Protesters

The Whale and Dolphin Coalition, made up of Jonny Lewis's crew from the studio in Sydney, morphed into **Greenpeace Australia**. Richard Jones registered the name and raised funds. A Sydney journalist, Jodi Adams, became the first coordinator and did the hard work of building an organisation. Assets included a Zodiac and equipment from the Albany campaign, the first direct action in Australia under the Greenpeace banner. In 2007, Greenpeace celebrated thirty years in Australia, dating from 28 August 1977, the day of the protest at the gates of the Cheynes Beach Whaling Company.

Tom Barber and **Aline Charney Barber** live in Reno, Nevada, where they maintain their practical approach to making a difference. Tom designed the world's first commercial wind farm, a positive alternative to nuclear energy. They used the lessons learnt in Albany to make the Californian wind farm successful. Tom designs furniture, things of beauty, and Aline works on green projects and campaigns to maintain Nevada's open spaces.

'People are part of the equation,' says Aline. 'Saving whales is the same as saving people. All people are entitled to a living. They have families, children and livelihoods. You can't come into a town and tell the community they're horrible. You have to come up with solutions, find people-alternatives to whaling, such as fishing or whale watching.'

The couple keeps in touch with Jonny Lewis, who is godfather to their son, Tatlin. Their older son, Ulysses, lives in Sydney. Tom still reckons Jonny is a dolphin, almost.

Tom says, 'The protest was about love and peace. Whales had a beauty and it was about preserving that beauty.' In the right light, you can still see the sunburn scars on Tom's face from exposure in an open boat off the Australian continental shelf.

Dr Peter 'Bro' Brotherton, who ran the early community campaign in Western Australia against Cheynes Beach Whaling Company, lives in Melbourne, where he has a consultancy called

Sustainable Solutions. He is a climate change campaigner and was a vice-president of the Australian Conservation Foundation.

Pat Farrington, in the decades since the campaign, drew strength from Bob Hunter's words to her as she departed Albany. 'I will be able to tell my grandchildren I tried,' he said. Pat lives in California and still feels a connection with the whales and dolphins.

'I felt for the townspeople of Albany most of all, knowing the struggle they had ahead of them,' Pat says. 'Knowing it might mean that even more of their young people and families would move away. Knowing how pride and care of the family was important to some of the men who would lose their jobs. Knowing the town's revenue base could shift disastrously and erode maintenance of the town's infrastructure and delivery of services just when suffering families and residents needed support the most.'

Tony Gregory in Sydney and **Henrietta Kaye** in Melbourne, who worked tirelessly with the Project Jonah team, maintain their connection to whales. Tony keeps getting invitations to speak about whales. Henrietta is an adviser to Project Jonah. She was made a Member of the Order of Australia for her work through Project Jonah to protect whales and dolphins. **Joy Lee** was also made a Member of the Order of Australia.

Bob Hunter continued as an advocate for the planet. He saw the creation of Greenpeace International, the global Rainbow Warriors. Hollywood still flirts with making a blockbuster about the wild defenders of the ecosphere; Bob helped write the script. The question of who would play him hasn't yet been decided. Bobbi always wanted her husband to write a book specifically about the Australian protests but there was always another project with greater priority. He did write about the Australian campaign in a section of his book, *The Greenpeace Chronicle* (*Warriors of the Rainbow* in North America).

Bob died on 2 May 2005, aged sixty-three. During the final years of his life he concentrated on creating public awareness of global warming. He wrote the book *2030: Confronting Thermageddon in Our Lifetime*. Thermageddon is a marriage of the words thermal and

Armageddon, the final battle between good and evil. He wrote thirteen books, bashed out countless pieces of journalism, lectured, had his own TV spot, and documented the living history of eco-defenders. He tried his hand at gaining a seat in the Canadian Parliament but lost that battle. He did, however, inspire others to enter politics and they continue to follow his ideals. He was awarded the Canadian Governor General's Award for Non-Fiction in 1991 for his book, *Occupied Canada: A Young White Man Discovers his Unsuspected Past*. In 2000 *Time Magazine* listed him as one of the Heroes of the Century for his efforts to defend nature. He received many honours after his death. The Bob Hunter Memorial Park, 550 acres at Markham, Ontario, was named in his honour by the Ontario Government.

Bob Hunter said, 'I have visited thirty-four countries in my journeys around the world, have swum underwater with dolphins, stood on ice in the path of an onrushing icebreaker, parachute-jumped, dodged great white sharks and motorcycle gangs of whale factory workers in Australia, faced angry mobs in Newfoundland, founded a religion, run with the bulls in Pamplona, survived numerous storms and other near-disasters at sea while commanding a converted minesweeper in the North Pacific, stuck my head in a killer whale's mouth and have nearly drowned or been stomped, run down or crushed many, many times … what a fabulous existence.'

Emily Hunter, Bob's daughter, scattered her father's ashes on an iceberg in the Antarctic on 6 January 2006. It was a brief interlude before Emily and her father's old friend Paul Watson, founder of the Sea Shepherd Conservation Society, continued chasing down the Japanese Government's whaling fleet.

Bobbi Hunter was a committed partner to Bob for thirty-one years. She continues to promote wise energy use through political efforts. Bobbi says, 'In 1977 there wasn't such a thing as a professional protester. Most strategy was done after you left the office for the day and hit the pub, drinking beers until the ideas flowed. We were flying by the seat of our pants. The impression was that Greenpeace was this large organisation but, in reality, it was twenty people from Vancouver.'

In 2007, Bobbi and Pat Farrington travelled together to Albany for the first time in thirty years. A reception was held for them at the offices of the City of Albany. Bobbi went whale-watching, met with members of the Albany Whale Protection Group and had a drink with Kase Van Der Gaag.

Richard Jones, Fund for Animals founder, Project Jonah funding wizard, Greenpeace Australia co-founder, and publisher of *Simply Living* magazine, was a member of the Legislative Council in New South Wales for almost fifteen years until 2003.

'You might say that this whole campaign was an absolute classic,' Richard says. 'No one thing won it. Not the action at Cheynes Beach, not the tens of thousands of signatures, not the ads, not the lobbying. It took all of those together to build up huge public, then political, support. This campaign should serve as a blueprint on how to win an issue. It not only stopped Australia whaling, but turned Australia from being one of the most fervent pro-whaling countries into the most anti-whaling country. And that remains true today, thirty years later.'

Jonny Lewis, Greenpeace Australia co-founder, says: 'The campaign took about a year and a half out of our lives — a long time for a young person. I wanted to get on with my life and I was happy with the outcome, thrilled. We created a lot of noise and we risked our lives doing it. That was the exciting part. We did our little bit and there were lots of other people who worked hard. I get emotional with every whale sighting on the east coast.'

Jonny regularly exhibits his photographs, including many from regional hot spots such as Bougainville and East Timor. His photographs are in public and private collections in Australia and around the world. He lives in a solar-powered house in the Southern Highlands of New South Wales, where he continues his interest in protest movements, the environment and indigenous issues.

Jonny Lewis, Kase Van der Gaag and Paddy Hart shook hands for the first time in Albany in 2007. They stood on Middleton Beach, where thirty years earlier Jonny and his friends launched their Zodiacs, and together led a protest against plans by Japan to take fifty humpback whales. Japan eventually postponed those plans, citing Australia's curious habit of giving names to humpbacks, but continued its annual program of catching minke whales under what Japan calls a scientific whaling program.

The foes of the anti-whaling campaign meet for the first time at Middleton Beach, Albany, where Greenpeace launched its Zodiacs in 1977. They came together on 3 November 2007 to protest plans by Japan to take fifty humpback whales. From left: Kase Van Der Gaag, former whaling ship master and gunner; Jonny Lewis, anti-whaling activist; Paddy Hart, former whaling ship master and gunner.

Photograph by Ken Matts. Copyright IFAW.

Albany

The Esplanade Hotel, headquarters of the 1977 activists, was rebuilt and at last report was about to be redeveloped again. No-one misses the old building and its view of the car park. The foreshore near the jetty, where the whale ships tied up, was also being redeveloped to include an entertainment centre.

Albany is a desirable place to live. More than thirty thousand people live in the town and more than fifty thousand in the region. There are two newspapers. New red-tiled suburbs have been built. Some residents are not happy at proposals to build multi-storey buildings in the historic precinct of Western Australia's oldest town, and there's even talk of installing traffic lights to handle the influx of cars during busy shopping hours.

The Port of Albany is Australia's largest, and the fifth in the world, exporter of woodchips made from farmed trees. Around fifteen per cent of Australia's grain is shipped from Albany.

Albany is a showcase for wind power. In 2001 a wind farm was built to provide seventy-five per cent of the town's electrical power needs. The twelve 1800kw wind turbines are sixty-five metres high with thirty-five-metre blades. They were the largest to be installed in the southern hemisphere.

The whale continues to play a defining role in Albany. The symbol can be seen throughout the town and whale-watching is an important part of the tourism industry. About 590,000 people come to the region each year on holiday. They spend $171 million.

Humpback whales, toothless cousins to sperm whales, are on the increase in Albany, as they are all around Australia. A whale-watching industry has grown up around the humpbacks' habit of coming close to shore. They come in close to Middleton Beach, where the Zodiacs were launched in 1977. Depending on the weather and number of whales making an appearance in any given year, up to four tour operators take tourists into Albany's King George Sound.

At last count, 1.6 million people spend a collective $300 million to see whales in Australia each year.

At the Natural Bridge, on the way to Whale World, there is a plaque:

ALL GOES IF COURAGE GOES (James Barrie)

On the night of March 14th, 1978, Stephen Matthews was saved from almost certain death in the waters off this spot. This plaque commemorates the outstanding courage of Paddy Hart, skipper, Keith Richardson, mate, and the crew of the whale chaser Cheynes II *and John Bell, aircraft pilot, in making the rescue.*

The Ships

Cheynes II, the last registered steamship in Western Australia, went to Tasmania, came home, went to Heard Island, came home under sail after running out of fuel, and was later fitted out as a restaurant. It is now a rusting hulk run aground in Princess Royal Harbour in Albany.

Cheynes III was sunk off Michaelmas Island in King George Sound as a recreational scuba diving site. Its three-phase steam engine is at the whaling station.

Cheynes IV is on dry land at the whaling station museum complex. Tens of thousands of people clamber over it every year. The harpoon gun deck is popular for photographs.

The Whales

The sperm whales hunted by the Cheynes Beach Whaling Company are elusive. Their health, in terms of numbers, is unknown and difficult to assess. It's a long and uncomfortable trip to the continental shelf and there's no guarantee of sighting sperm whales, hardly an attractive tourism project. It's safe to assume these toothed whales follow their ancient migratory path along the bottom of Australia each year, as they have for thousands of years, hunting giant squid in the undersea canyons along the continental shelf.

The Albany whalers have spent more time observing Southern Ocean sperm whales than anyone alive today. They are convinced there are a bloody lot of whales out there. The whalers know the sperm whale pods will be splitting off to create new pods. Former bachelor males will be creating their own harems and these pods in turn will grow until they break to form new groups.

It is debatable whether these sperm whales, the population group called Division 5 by the International Whaling Commission, have regained their natural equilibrium in terms of numbers, even after three decades. A female sperm whale reaches maximum size after twenty years. A male takes thirty years and may not mate until he's in his mid-twenties.

When scientists at the International Whaling Commission in 1978 recalculated population sizes for the sperm whales off Albany, they found numbers had dropped so low that the only recommendation could be a zero catch limit. By that time, the Cheynes Beach Whaling Company had already decided to close. The company knew what was coming. Close contact was kept with the scientists of the commission.

The serious problem was male sperm whales, the sex favoured as a harpoon target. In 1978, when whaling ended, numbers of male sperm whales had fallen to 4,700, down from 18,200 in 1947. The female population was in better shape at 24,300, down from 26,700.

These calculations do not take into account the impact whaling in the nineteenth century may have had on this sperm whale population. The figures are based on sound theory but they are only guesses. No-one really knows.

The big question: were there enough physically and socially mature males to rebuild the Albany sperm whale population? It is likely that pregnancy rates were in steep decline in the last few years of Australian whaling because there were so few males. Again, this is shown through statistical modelling. It is also likely that the numbers of females fell substantially in the years following the end of whaling. There weren't enough males and there still may not be enough. One set of figures suggests the Albany sperm whales won't recover until somewhere between 2024 and 2029.

Big male sperm whales are swimming free in the Southern Ocean. Without culling, the average size and age of the sperm whales off Albany will have increased. Some grow to more than twenty metres and the sperm whale is the only animal in the world physically capable of swallowing an adult human whole. The male newborns of 1979 reached sexual and social maturity in the last few years and may now have their own harems.

With a lifespan of sixty years or more, there will be sperm whales alive today that have experienced, and perhaps still remember, being hunted off Albany, Western Australia.

Postscript : The Phantom

The anti-whaling activist known as Jean-Paul Fortom-Gouin left Australia after the 1977 campaign, but the Phantom always reappeared throughout the world when his money and expertise were needed to free the whales. He was born Paul Gouin but extended this to create Jean-Paul Fortom-Gouin. Fortom sounds like 'Strong Man' in French. He spent about $100,000 dollars on the Australian campaign, and he can't be sure but believes he gave around one million dollars to the global campaign for the whales.

'I went fifty-fifty, half for myself and half for the whales,' he says. 'I was worth maybe two million dollars in those days. It was my money and I was happy to spend it. I was making my money at that time in asset management. I had clients who let me manage their money in a tax haven, and if I got them more than a set percentage I got a significant amount. I was either pretty good at it or pretty lucky.'

Jean-Paul is amused that people saw him as a smart dresser and a powerful, attractive figure. 'I guess it could have been the aura of the guy writing the big cheques. I have never cared about the way I dress and no-one has ever complimented me. I am a tightwad when it comes to personal expenses.' The whales were his passion and his indulgence. He found his suits in thrift shops.

'It was an adventure. It's a story of a small group of dedicated people who had a bug up their ass and they went out and did something about it. You *can* make a difference. The fact that Australia switched camp was very influential to the whole picture.'

He believes the Albany campaign fulfilled its purpose — to get the whales on to the front pages — but Project Jonah was more influential in the long run. 'They were doing it more deeply than us. They were going to the schools.'

He kept his role as commissioner for Panama and each year he nominated himself to the scientific committee. He wrote more papers, including one arguing that no female Cachalot should be hunted. These whales are similar to elephants because they have matriarchal harems controlled by large females and one large male. Taking out the larger females removed the cohesion of the whale pods and the family disintegrated. But the accepted theory at that time was that the more you killed them, the faster they bred.

Jean-Paul was behind the pelagic (open ocean) whaling moratorium in 1980. As the Panamanian commissioner, he proposed a moratorium before the main meeting. However, he added up the numbers and realised there was no way he could get the three-quarter majority vote needed for a successful motion. But if he split the vote between coastal and pelagic whaling, there was a chance to stop pelagic whaling because there were only two countries doing it — Japan and USSR.

'So there might be a way to put pressure on the coastal whaling countries to either abstain or vote against pelagic,' he says. 'I proposed that during the meeting to the chairman, Asgeirsson, the Icelandic commissioner.'

Asgeirsson told him the motion was a major modification to the meeting agenda and it should have been filed sixty days before.

'I had read the rules and regulations and you could challenge the chairman if you didn't like his ruling. I challenged it. It went to a vote and I lost it. When I looked at what happened, the French delegate had misunderstood the vote. The French in fact wanted to split the vote. I asked for another vote. The chairman said it would be chaos if any delegate could ask for another vote if he had misunderstood the issue. He had a good point but I challenged his ruling anyway. We had another vote and by then everybody knew how they should vote. So we won the vote whether to have another vote, and then another vote to split the proposal.'

The behind-the-scene negotiations started. The only way to get the vote through was to exclude minke whales from the pelagic moratorium.

'We got it passed without the minke whales. This meant several thousand Cachalot were saved. They are my totem animals: they are sacred to me; they have the biggest and most complex brains on earth; they are awesome; they are my personal responsibility. I managed to get the Cachalot pelagic whaling stopped. I can say that I saved thousands of Cachalot from being hunted by being at the right place and having the right idea.'

Following that 1980 meeting, Jean-Paul moved to the second stage, a general moratorium on whaling. Peter Scott, the president of World Wildlife Fund, invited Jean-Paul and David McTaggart, then the president of Greenpeace International, for a strategic meeting at his home in England. 'We decided to go looking for new IWC members.'

Latin America and the Caribbean was Jean-Paul's beat. He went to work convincing non-IWC countries to join and vote with the anti-whaling lobby. With the help of Dr Francisco Palacio, a professor at the Rosenstiel School of Marine Science in Miami, he brought four countries into the commission: Antigua, Belize, Costa Rica and St Vincent.

Again, a three-quarter majority vote is needed for a moratorium. It was passed twenty-five votes to seven, with Spain abstaining because they saw that the moratorium was going to pass even if they voted no.

'If I hadn't brought those four countries, the general moratorium would not have passed that year for sure. And maybe it would never have happened because Japan soon started to bring in countries as well. It is very difficult to get a three-quarter majority.'

Eventually pressure from Japan saw Jean-Paul lose his role as commissioner for Panama. 'We never had to go and bribe a country. We went and convinced them to be with the good guys. Some times the little countries had to be helped by some NGO (non-government organisation) to pay their fees, but we never bribed them. They never made money out of it, just some expenses paid, such as the IWC fees. We told them, "If you want to be with the good guys, vote with England; vote with the US; vote with France".'

Jean-Paul's last success was in 1994. The idea was to have a Southern Ocean sanctuary for whales. This had no scientific basis. Everyone agreed there were lots of minke whales in the Antarctic Ocean and they certainly were not close to extinction.

'It was worth giving it a shot,' he said. 'I thought it would close Japanese pelagic whaling because the Antarctic minke were the last abundant population of whales. But the Japanese Government keeps it ticking along with phoney "scientific" permits, which are denounced every year.'

He couldn't put the proposal to the IWC because he was no longer an official delegate. He contacted different NGOs in France to lobby the French Government but nothing happened. David McTaggart knew the French Minister for the Environment, Brice Lalonde, and gave him a call. Lalonde, a Green Party leader, told him he had resigned that day and that he was clearing his desk. David told him to dictate a letter to the IWC before leaving his position that France wanted to put the Antarctic Sanctuary on the agenda of the next meeting.

Four years later in Puerto Vallarta, Mexico, the Southern Ocean Sanctuary went to the vote. Jean-Paul says, 'That was one of the big joys of my life, to witness this vote, and to think it all started with one idea. Only Japan voted against. The smaller countries abstained. That sanctuary was the work of many, many people. And without those people it would never have happened. But the original idea was mine. I proposed it to the NGOs at the 1990 IWC meeting at Noordwijkerhout in Holland. They didn't believe then it could possibly succeed.'

Jean-Paul handed control of the Florida-based Institute for Delphinid Research to the two trainers who ran it. He stopped going to whaling commission meetings. He lost contact with anti-whaling groups. 'I figured things were pretty much secure so I semi-retired from the whaling issue.'

The Phantom melted back into the jungle.

EMAIL

To: Jonny Lewis
Subject: From the Phantom
Date: Sunday 25 March 2007

Hola, amigo,

A ghost from the past, AKA the Phantom.

Very much alive and having a good time as always.

I haven't done much for the whales after the two moratoria and the South Ocean Sanctuary because I felt the situation more or less under control.

But it looks like the bad guys are creeping back and it is time for the next initiative.

I am sure that you want to be part of it.

Jean-Paul

Greenpeace founders' night in September 2007. Veterans of the 1977 Albany campaign, the first direct action by Greenpeace in Australia, gather for the first time in thirty years. From left: Jodi Adams, the first coordinator of Greenpeace in Australia; Richard Jones who registered Greenpeace in Australia; Stephen Jones who took part in the Albany campaign; Tom Barber who piloted a Zodiac in Albany; Chris Pash who covered the action for the *Albany Advertiser*; Jean-Paul Fortom-Gouin who financed the campaign and piloted a Zodiac; Jonny Lewis who formed the Whale and Dolphin Coalition and piloted a

Zodiac; American Pat Rose Farrington who played a key role at the whaling station protest; Canadian Bobbi Hunter, Greenpeace's first treasurer and a member of the 1977 Albany team; American Aline Charney Barber, a member of the direct action team in Albany. Missing: the late Bob Hunter; Australian Allan Simmons, Albany campaigner and co-mechanic with Tom Barber on the truck and Zodiac motors.

Photograph by James Alcock, Copyright Greenpeace.

Acknowledgements

This story was created with the voices, thoughts, actions and feelings of those who experienced the end of whaling in Australia. I was there and floated around the edges as an observer but the story belongs to those who played a role.

Many helped bring this work to reality and many more endured my discussion of *The Last Whale*'s progress as I dug deeper, sorting fact from myth.

Here are some of them: Ed Smidt, photographer, colleague and friend, whose knowledge of Albany, Western Australia, and its people is bottomless and who shared adventures in the 1970s; manuscript reader, sanity checker and supporter Ian Williams of Adelaide; editor, mentor and writer Tom Flood; the team at Fremantle Press for their professionalism and vision.

I started speaking to some of the people in this book more than thirty years ago. They displayed considerable honesty, patience and faith as they answered questions and drew detail from memory.

Here are some of them: Aline Charney Barber; Pat Rose Farrington; Bobbi Hunter; Emily Hunter; Jonny Lewis; Jean-Paul Fortom-Gouin; Tom Barber; Richard Jones; Henrietta Kaye; Tony Gregory; Peter 'Bro' Brotherton; Kase Van Der Gaag; Paddy Hart; Mick Stubbs; the late Gordon Cruickshank; the late John Bell; the late Ches Stubbs; Peter 'PJ' Johnson; John Saleeba; Keith Forde; Barbara Cruickshank; Stephen Matthews; former prime minister Malcolm Fraser; Phoebe Fraser.

A special mention to Aline Charney Barber who shared her personal diary of 1977.

On the documentation side: Peter Canole, Historian, Western

Australian Police; Yvonne Wallace, Curator, Whale World; the Battye Library oral history collection; Western Australian State Archives; the National Library of Australia.

The narrative was assisted by oral histories. I commend the Albany whaling oral history project and the interviews by Yvonne Choules.

Chris Puplick, former senator, who shared his memories and gave access to his paper *Inside the Whale: A History of Australia's Policy on International Whaling* (Sydney, January 1997).

Rex Weyler, author of *Greenpeace: How a group of Ecologists, Journalists, and Visionaries Changed the World*, was generous with information and in helping find people.

Photographs: Ed Smidt; Jonny Lewis; Aline Charney Barber; *The West Australian*.

And to my wife, Lee Rushton, and children, Amelia and Augustus, all of whom knew the answer before they asked: 'Where's dad?'

Working on *The Last Whale*.

Chris Pash, former editor, correspondent, bureau chief and newswire chief executive, was a pimply reporter at the *Albany Advertiser* in 1977 when activists launched Greenpeace's first direct action in Australia.

He now lives in Sydney, Australia, with his wife and two children and is an executive in the news and information industry.

For further information on Greenpeace Australia visit:

www.greenpeace.org/australia

For the latest information on the campaign to save the whales visit Chris Pash's site:

http://thelastwhale.blogspot.com

For information on other Fremantle Press titles see:

www.fremantlepress.com.au

First published 2008 by
FREMANTLE PRESS
25 Quarry Street, Fremantle
(PO Box 158, North Fremantle 6159)
Western Australia.
www.fremantlepress.com.au

Copyright © Chris Pash, 2008.
Copyright in the photographs remains with the individual photographers.

This book is copyright. Apart from any fair dealing for the purpose of private study, research, criticism or review, as permitted under the Copyright Act, no part may be reproduced by any process without written permission. Enquiries should be made to the publisher.

Consultant editors Ray Coffey and Wendy Jenkins
Designer Tracey Gibbs
Cover and frontispiece photo: MINDEN PICTURES/Emerald City Images
Printed by Everbest Printing Company, China.
Production specification: paper 100% recycled; print Certification ISO 14001; inks vegetable-based; water-based varnish on the cover.

National Library of Australia
Cataloguing-in-publication data

Pash, Chris.
The last whale / Chris Pash.

North Fremantle, W.A. : Fremantle Press, 2008.
9781921361326 (pbk.).

1. Greenpeace Australia. 2. Whales — Conservation — Australia.
3. Whaling — Australia.
333.9595160994

Publication of this title was assisted by the Commonwealth Government through the Australia Council, its arts funding and advisory body.